"From fears to phobias and panic, this [...] mysteries of worry, then delivers strate[...] resilient life."

—**Reid Wilson, PhD**, author of *Don't Panic*

"Fear, worry, anxiety, panic, and depression block individuals from living fully. This book does an excellent job of unraveling the mysteries of the brain and how it causes anxiety, thus allowing individuals to take control of their symptoms and manage their lives more effectively. I have used the concepts in this book in treating veterans with PTSD and have seen people become much more functional. Highly-recommended reading."

—**Susan Myers, RN, LCSW, BCD**, holistic nurse and board certified diplomate in clinical social work

"This is a unique contribution to a wide array of self-help books for those who suffer with anxiety. The authors explain what we know about the workings of the brain in a fluid style that neither talks down to nor overwhelms the reader. This science becomes the foundation for decreasing bewilderment, fear, and shame. There are straightforward and logical recommendations for modifying patterns of anxiety that originate from, and are maintained by, differing brain circuitry."

—**Sally Winston PsyD**, codirector of the Anxiety and Stress Disorders Institute of Maryland

"Catherine Pittman, a trained behavioral scientist, brings her deep scientific understanding of fear, anxiety, and learning to the world of personal experiences. Few scientists can make this transition from scientific communication to public communication. Pittman, along with coauthor Elizabeth Karle, does this very well indeed. Readers should not be put off by the early presentations of brain mechanisms, because they are surprisingly readable and informative. Moreover, it is information that provides a foundation for readers who have anxiety challenges to use later as they develop effective coping strategies. Readers should find the clear expositions of the where's, why's, and how's of anxiety and its management to be an anxiety-reducing read."

—J. Bruce Overmier, PhD, professor emeritus in the graduate faculties of psychology, neuroscience, and cognitive science at the University of Minnesota

Rewire
Your
Anxious
Brain

how to use the neuroscience of fear to end anxiety, panic & worry

Catherine M. Pittman, PhD
Elizabeth M. Karle, MLIS

New Harbinger Publications, Inc.

Publisher's Note

This publication is designed to provide accurate and authoritative information in regard to the subject matter covered. It is sold with the understanding that the publisher is not engaged in rendering psychological, financial, legal, or other professional services. If expert assistance or counseling is needed, the services of a competent professional should be sought.

NEW HARBINGER PUBLICATIONS is a registered trademark of New Harbinger Publications, Inc.

Distributed in Canada by Raincoast Books

Copyright © 2015 by Catherine M. Pittman & Elizabeth M. Karle
New Harbinger Publications, Inc.
5720 Shattuck Avenue
Oakland, CA 94609
http://www.newharbinger.com

Cover design by Amy Shoup
Acquired by Jess O'Brien
Edited by Jasmine Star

Library of Congress Cataloging-in-Publication Data

Pittman, Catherine M.
  Rewire your anxious brain : how to use the neuroscience of fear to end anxiety, panic, and worry / Catherine M Pittman and Elizabeth M Karle.
    pages cm
  Includes bibliographical references.
  ISBN 978-1-62625-113-7 (paperback) -- ISBN 978-1-62625-114-4 (pdf e-book) -- ISBN 978-1-62625-115-1 (epub) 1. Anxiety--Prevention. 2. Worry--Prevention. 3. Fear. 4. Neuroplasticity. 5. Neuropsychology. I. Karle, Elizabeth M. II. Title.
  RC531.P47 2015
  616.85'2205--dc23
                          2014046563

Printed in the United States of America

26   25   24

25   24   23   22

This book is dedicated to all of the children and adults who suffer from anxiety or panic, and who need daily courage to find their way through the experience. We hope this book can help them to live the lives they wish for themselves.

Part 1
Anxious Brain Basics

Part 2
Taking Control of Your Amygdala-Based Anxiety

Part 3
Taking Control of Your Cortex-Based Anxiety

Acknowledgments

My work on this book would not have been possible without the assistance and support of many people in my life, and I would like to thank them here.

First, of course, my coauthor and partner, Elizabeth (Lisa) Karle, who has enriched my life in countless ways and accompanied me into a variety of endeavors I could never have imagined without her. She amazes me on a daily basis with her courage in the face of her own anxiety, her patience with all that life requires of her, and her determination to hold herself to high standards.

My daughters, Arrianna and Melinda, who have tolerated months of me working on my laptop, not to mention years of discussion of the amygdala and cortex. I hope they know how much I love them despite many evenings of researching and writing.

My clients over the past thirty years, who have taught me so much and inspired my respect and admiration as they've retrained their brains, shaping their lives to follow their dreams. They haven't let their struggles with anxiety or brain injury keep them from becoming who they were meant to be.

William (Bill) Youngs, neuropsychologist and dear friend, who has provided a wealth of knowledge and encouragement during our weekly lunches over the past twenty-five years, and who made many valued observations and suggestions during the creation of this book.

Cathy Baumgartner, administrative assistant and friend, who's made the Psychology Department run smoothly while I've served as chair and who made it possible for me to spend precious hours in the library during recent months. I feel so fortunate to have her competence and sense of fun in my life.

Samantha Marley, a psychology major at Saint Mary's and student assistant in the Psychology Department, who helped not only by scoring exams but also by working on the many references for this book. After her senior thesis, Sam produces perfectly formatted references in no time!

—Catherine

Having a mental illness of any kind is a challenge. Not only can it affect daily living, but it can also alter the trajectory of one's life plans. Often it doesn't stop there, as the ups and downs of anxiety and other disorders impact family, friends, and coworkers as well. We hope this book will provide insights and information that will help our readers and their support systems weather these challenges. We are grateful to the professionals at New Harbinger Publications for giving us the opportunity to share our knowledge and experience with you.

On a personal note, I'd like to thank the members of my own support system for always being there: my parents and siblings, whose love knows no bounds; Carol, who amazes me; Brother Sage, for his daily wit and wisdom; Janet and my colleagues at Saint Mary's College, for their patience and assistance; Tonilynn, who understands better than anyone else; Bill, the brain master; my ggf Guiseppe Carpani, for being in the right place at the right time; and, of course, Catherine, with whom I've shared both meaningful dreams and madcap adventures.

Lastly, a special thank-you to my nieces and nephews, whose limitless joy and affection make the sights and sounds of life more rewarding. "To infinity and beyond!"

—Elizabeth

INTRODUCTION

The Pathways of Anxiety

You're driving to work one day when you suddenly wonder, *Did I turn off the stove?* You begin to mentally trace your steps from earlier that morning, but you still can't remember turning it off. You probably did...but what if you didn't? Your anxiety begins to build as the image of the stove catching on fire pops into your head. Just then, the person in the car in front of you slams on the brakes. You clutch the steering wheel tightly and hit your own brakes hard, stopping just in time. Your whole body is activated with a surge of energy and your heart is pounding, but you're safe. You take some deep breaths. That was close!

Anxiety, it seems, is all around us. If you carefully consider the events in the scenario above, you'll notice that they illustrate two very different ways that anxiety begins: through what we think about, and through reactions to our environment. This is because anxiety can be initiated by two very different areas of the human brain: the cortex and the amygdala. This understanding is the result of years of research in a field known as *neuroscience*, which is the science of the structure and function of the nervous system, including the brain.

The simple example above, involving both the imagined stove and the braking car, illustrates the underlying principle of this book: two separate pathways in the brain can give rise to anxiety, and each pathway needs to be understood and treated for maximum relief (Ochsner et al. 2009). In that example, anxiety was aroused in the cortex pathway by thoughts and images of the risks of leaving the stove on all day. And information from another anxiety-producing pathway, traveling more directly through the amygdala, ensured a quick reaction to avoid rear-ending another car.

Everyone is capable of experiencing anxiety through both pathways. Some people may find that their anxiety arises more frequently in one pathway than the other. As you'll learn, recognizing the two pathways and handling each in the most effective manner is essential. The purpose of this book is to explain the differences between the two pathways, demonstrate how anxiety is created in each, and give you practical ways to modify circuits in each pathway in order to make anxiety less of a burden in your life. We'll show you how you can actually change the pathways in your brain so that they're less likely to create anxiety.

Understanding Anxiety

Anxiety is a complex emotional response that's similar to fear. Both arise from similar brain processes and cause similar physiological and behavioral reactions; both originate in portions of the brain designed to help all animals deal with danger. Fear and anxiety differ, however, in that *fear* is typically associated with a clear, present, and identifiable threat, whereas *anxiety* occurs in the absence of immediate peril. In other words, we feel fear when we actually are in trouble—like when a truck crosses the center line and heads toward us. We feel anxiety when we have a sense of dread or discomfort but aren't, at that moment, in danger.

Everyone experiences fear and anxiety. Events can cause us to feel in danger, such as when a severe storm shakes our house or when we see a strange dog bounding toward us. Anxiety arises when we worry about the safety of a loved one who's far from home, when we hear a strange noise late at night, or when we contemplate everything we need to complete before an upcoming deadline at work or school. Many people feel anxious quite often, especially when under some kind of stress. Problems begin, however, when anxiety interferes with important aspects of our lives. In that case, we need to get a handle on our anxiety and regain control. We need to understand how to deal with it so it no longer limits our lives.

Anxiety can limit people's lives in surprising ways—many of which may not seem to be due to anxiety. For example, while some people are plagued by worries that haunt every waking moment, others may find it difficult to fall asleep. Some may have a hard time leaving home, while for other individuals a fear of public speaking may threaten their job. A new mother may have to complete a series of rituals for hours each morning before she can leave her child with a sitter. A teenage boy may be haunted by nightmares and get suspended for fighting in school after his home has been destroyed by a tornado. A plumber's anxiety about encountering large spiders may reduce his income to a level that won't support his family. A child may be reluctant to attend school and unwilling to talk to her teachers, threatening her education.

Even though anxiety has the power to rob a person of the capacity to complete many of the basic activities of life, all of these individuals can return to fully engaging in life. They can understand the cause of their difficulties and begin to find confidence again. This understanding is possible thanks to a recent revolution in knowledge about the brain structures that create anxiety.

In the past two decades, research on the neurological underpinnings of anxiety has been conducted in a variety of laboratories around the world (Dias et al. 2013). Research on animals has uncovered new details about the neurological foundations of fear. Structures in the brain that detect threats and initiate protective responses have been identified. At the same time, new technologies like functional magnetic resonance imaging and positron emission tomography scans have provided detailed information about how the human brain responds in a variety of situations. When reviewed, analyzed, and combined, this emerging knowledge allows neuroscientists to make connections between animal research and human research. As a result, they are now able to assemble a clear picture of the causes of fear and anxiety, providing an understanding that surpasses our understanding of all other human emotions.

This research has revealed something very important: two fairly separate pathways in the brain can create anxiety. One path begins in the *cerebral cortex*, the large, convoluted, gray part of the brain, and involves our perceptions and thoughts about situations. The other travels more directly through the *amygdalas* (uh-MIG-dull-uhs), two small, almond-shaped structures, one on each side of the brain. The amygdala (generally referred to in the singular) triggers the ancient fight-or-flight response, which has been passed down virtually unchanged from the earliest vertebrates on earth.

Both pathways play a role in anxiety, although some types of anxiety are more associated with the cortex, while others can be directly attributed to the amygdala. In psychotherapy for anxiety, attention has typically been focused on the cortex pathway, using therapeutic approaches that involve changing thoughts and arguing logically against anxiety. However, a growing body of research suggests that the role of the amygdala must also be understood to develop a more complete picture of how anxiety is created and how it can be controlled. In this book, we'll explore both pathways to give you a full picture of anxiety and how to change it, whatever its origin.

The Cortex and the Amygdala

Chances are you're already familiar with the cortex, the portion of the brain that fills the topmost part of the skull. It's the thinking part of the brain, and some say it's the portion of the brain that makes us human because it enables us to reason, create language, and engage in complicated thinking, such as logic and mathematics. Species that have a large cerebral cortex are often thought to be more intelligent than other animals.

Approaches to treating anxiety that target the cortex pathway are numerous and typically focus on *cognitions*, the psychological term for the mental processes that most people refer to as "thinking." Thoughts originating in the cortex may be the cause of anxiety, or they may have the effect of increasing or decreasing anxiety. In many instances, changing our thoughts can help us prevent our cognitive processes from initiating or contributing to anxiety.

Until recently, treatments for anxiety were less likely to take the amygdala pathway into consideration. The amygdala is small, but it's made up of thousands of circuits of cells dedicated to different purposes. These circuits influence love, bonding, sexual behavior, anger, aggression, and fear. The role of the amygdala is to attach emotional significance to situations or objects and to form *emotional memories*. Those emotions and emotional memories can be positive or negative. In this book, we'll focus on the way the amygdala attaches anxiety to experiences and creates anxiety-producing memories. This will help you to understand the amygdala so you can learn how to change its circuitry to minimize anxiety.

We humans aren't consciously aware of the way the amygdala attaches anxiety to situations or objects, just as we aren't consciously aware of the liver aiding digestion. However, the amygdala's emotional processing has profound effects on our behavior. As we'll discuss in this book, the amygdala is at the very heart of where the anxiety response is produced. Although the cortex can initiate or contribute to anxiety, the amygdala is required to trigger the anxiety response. This is why a thorough approach to addressing anxiety requires dealing with both the cortex pathway and the amygdala pathway.

The chapters in part 1 of this book, "Anxious Brain Basics," are dedicated to explaining the cortex and amygdala pathways. We'll explain the different ways the pathways work, both separately and in conjunction with one another. Once you have a good foundation in how each pathway creates or enhances anxiety, we'll teach you specific strategies to combat, interrupt, or inhibit your anxiety based upon what you've learned about the circuitry in your brain. We'll describe strategies you can use to change the amygdala pathway in part 2, and strategies to change the cortex pathway in part 3. Then, in the conclusion, "Putting It All Together," we'll help you draw upon everything you've learned about changing your brain in order to live a more anxiety-resistant life.

The Promise of Neuroplasticity

In the past two decades, research has revealed that the brain has a surprising level of *neuroplasticity*, meaning an ability to change its structures and reorganize its patterns of reacting. Even parts of the brain that were once thought impossible to change in adults are capable of being modified, revealing that the brain actually has an amazing capacity to change (Pascual-Leone et al. 2005). For example, people whose brains are damaged by strokes can be taught to use different parts of the brain to move their arms (Taub et al. 2006). Under certain circumstances, circuits in the brain that are used for vision can develop the capacity to respond to sound in just a few days (Pascual-Leone and Hamilton 2001).

New connections in the brain often develop in surprisingly simple ways: Exercise has been shown to promote widespread growth in brain cells (Cotman and Berchtold 2002). In some research, just *thinking* about taking certain actions, like throwing a ball or playing a song on the piano, can cause changes in the area of the brain that controls those movements (Pascual-Leone et al. 2005). In addition, certain medications promote growth and changes in circuits of the brain (Drew and Hen 2007), especially when combined with psychotherapy. Also, psychotherapy alone has been shown to produce changes (Linden 2006), reducing activation in one area and increasing it in others.

Clearly, the brain isn't fixed and unchangeable, as so many people, scientists included, once assumed. The circuits of your brain aren't determined completely by genetics; they're also shaped by your experiences and the way you think and behave. You can remodel your brain to respond differently, no matter what age you are. There are limits, but there's also a surprising level of flexibility and potential for change in your brain, including changing its tendency to create problematic levels of anxiety.

We'll help you use neuroplasticity, along with an understanding of how the cortex and amygdala pathways work, to make lasting changes in your brain. You can use this information to transform your brain's circuitry so that it resists anxiety, rather than creating it.

Don't Go It Alone

We strongly recommend that you seek professional help, and specifically cognitive behavioral therapy, as you work on the strategies presented in this book. Cognitive behavioral therapists are trained in identifying anxiety-producing thoughts and other techniques in this book, including exposure therapy. Therapists in many disciplines have training in cognitive behavioral therapy, including social workers, for example. When choosing a therapist, it's important to ask whether the therapist is knowledgeable about cognitive behavioral methods of treatment, especially exposure and cognitive restructuring.

If you take antianxiety medications, it's important to use them wisely to support the process of modifying your anxiety. If a family practitioner prescribes your medications, we strongly suggest that you consult with a psychiatrist, who will have more experience with antianxiety medications, and the brain and how medications affect it. In addition, psychiatrists are more likely to be familiar with exposure and cognitive behavioral therapy in general.

That said, psychiatrists aren't necessarily trained in the various amygdala-based and cortex-based strategies for reducing anxiety we outline in this book. Many people seeking treatment for anxiety expect a psychiatrist to provide therapy and are surprised when the psychiatrist instead focuses on medications. Remember, psychiatrists aren't therapists; they're physicians who are trained to treat psychological disorders, primarily through the use of medications.

If you speak with a psychiatrist about medications, be sure the two of you consider the distinction between medications that provide relief from anxiety on a short-term basis and those that can assist you in modifying your brain's anxiety responses in a more lasting way. Also, explain the approaches you're using to combat anxiety so that any medications you take support you in the process. And, of course, make sure to inform your psychiatrist about any medication side effects you experience. Good communication between you, your psychiatrist, and your therapist, if you have one, can be immensely helpful in facilitating the process of rewiring your anxious brain. Each of you can make important contributions in evaluating how a given medication is working and affecting the treatment process.

Understanding How Anxiety Limits You in Your Life

At its best, anxiety can help us stay alert and focused. It can get our hearts pounding and give us the extra adrenaline we need to, say, win a race. At its worst, however, it can wreak havoc with our lives and paralyze us to the point of inaction.

If you suffer from anxiety, especially an anxiety disorder, you know how disabling it can be. However, ridding yourself of all anxiety isn't a realistic goal; and it is not only impossible but unnecessary. For some people, fear of flying severely limits their career, but others can easily avoid air travel for an entire lifetime with few consequences. If you focus your attention on the anxiety reactions that frequently or severely interfere with your ability to live your life in the way you desire, you'll be on the right track.

Take some time right now to think of examples of how anxiety or avoidance interferes with your life. Write them down if that's helpful. Think of potential goals that you have difficulty accomplishing due to anxiety. And because anxiety can extend its reach to influence future decisions, be sure to look beyond your daily life. Is anxiety keeping you from doing things like taking a trip, switching jobs, or confronting a problem?

Of course, you can't tackle all of these situations at once. Several considerations are helpful in choosing which situations to focus on and which to focus on first. You could begin with the situations that you deal with most frequently, or you might want to start with situations that result in the highest levels of anxiety. In any case, it's essential to focus on situations in which reducing anxiety will make a real difference in your life.

Exercise: Identifying Your Life Goals

The central goal of this book is to give you the power to live your life in the manner you wish so that you can fulfill your own aspirations. Therefore, when deciding which anxiety responses you want to modify, carefully consider your personal objectives. What short-term and long-term goals do you have for yourself? To help you clarify this, complete the following sentences. For each sentence, try to imagine what you'd like to do if anxiety weren't a limiting factor:

In the future, I'd like to see myself…

In one year, I'd like to…

In eight weeks, I'd like to…

If I weren't so concerned about _____, I would…

Keeping in mind the anxiety responses that most impact your life, you're now ready to learn how to change those responses. So in chapter 1, we'll begin by taking a look at the two pathways in the brain that create anxiety. Learning how the circuitry in these pathways functions, along with how you can potentially bypass, interrupt, or change that circuitry, is the first step to changing your life.

PART 1

Anxious Brain Basics

CHAPTER 1

Anxiety in the Brain

We want to start this chapter with a promise that everything that we tell you about the brain in this book is useful, practical information that will illuminate the causes of anxiety and help you understand how to change your brain to decrease your experience of anxiety. We won't present detailed, technical descriptions of all the neurological processes involved; instead, we will provide a simplified, basic explanation of anxiety in the brain that can help you understand why certain strategies will help you control your anxiety.

Revisiting the Two Pathways to Anxiety

If you don't know what causes your anxiety, you're at a disadvantage when you try to change it. Anxiety is created by the brain, and wouldn't occur without the contributions of specific brain areas. And while the brain is a very complex, interconnected system, much of which remains a mystery, we can identify two general sources of anxiety in it. There are also techniques you can use to target these specific sources of anxiety that will help you be more effective in managing or preventing the anxiety you feel.

As mentioned in the introduction, the main sources of anxiety in the brain are two neural pathways that can initiate an anxiety response. The cortex pathway is the one most people think of when they consider the causes of anxiety. You'll learn much more about the human cerebral cortex in the next section. For now, we'll simply say that the cortex is the pathway of sensations, thoughts, logic, imagination, intuition, conscious memory, and planning. Anxiety treatment typically targets this pathway, probably because it's a more conscious pathway, meaning that we tend to be more aware of what's happening in this pathway and have more access to what this part of the brain is remembering and focusing on. If you find that your thoughts keep turning to ideas or images that increase your anxiety, or that you obsess over doubts, become preoccupied with worries, or get stuck in trying to think of solutions to problems, you're probably experiencing cortex-based anxiety.

The amygdala pathway, on the other hand, can create the powerful physical effects that anxiety has on the body. The amygdala's numerous connections to other parts of the brain allow it to mobilize a variety of bodily reactions very quickly. In less than a tenth of a second, the amygdala can provide a surge of adrenaline, increase blood pressure and heart rate, create muscle tension, and more. The amygdala pathway doesn't produce thoughts that you're aware of, and it operates more quickly than the cortex can. Therefore, it creates many aspects of an anxiety response without your conscious knowledge or control. If you feel like your anxiety has no apparent cause and doesn't make logical sense, you're usually experiencing the effects of anxiety arising from the amygdala pathway. Your awareness of the amygdala is likely to be based on your experience of its effects on you—namely bodily changes, nervousness, wanting to avoid a certain situation, or having aggressive impulses.

Therapists often don't discuss the amygdala when treating anxiety disorders, which is surprising, given that most experiences of fear, anxiety, or panic arise due to involvement of the amygdala. Even when the cortex is the source of anxious thinking, it's the amygdala that causes the physical sensations of anxiety to occur: pounding heart, perspiration, muscle tension, and so on. However, when family doctors and psychiatrists are prescribing medications to reduce anxiety, they're often focused on the amygdala, even though they may not mention it by name. These medications, such as Xanax (alprazolam), Ativan (lorazepam), and Klonopin (clonazepam), often have the effect of sedating the amygdala.

Such tranquilizing medications are very effective at quickly reducing anxiety. Unfortunately, they do nothing to change the circuitry of the amygdala. So while they reduce the anxiety response, they don't help change the amygdala in ways that would be beneficial in the long term. (If you're taking antianxiety medications or want to know how specific medications affect the process of treating anxiety, please visit http://www.newharbinger.com/31137, where you'll find a bonus chapter, "Medications and Your Anxious Brain," which is available for download. See the back of the book for information on how to access it.)

The amygdala has many functions that aren't related to anxiety, and we won't delve into them here. To understand the amygdala's role in anxiety, it's important to know that as you go about your day, the amygdala notices sounds, sights, and events even though you may not be consciously focused on them. The amygdala is on the lookout for anything that might indicate potential harm. If it detects potential danger, it sets off the fear response, an alarm in the body that protects us by preparing us to fight or flee.

Consider it this way: We are the descendants of frightened people. Early humans whose amygdala reacted to potential dangers and produced a strong fear response were most likely to behave in cautious ways and be protective of their children, which meant they were more likely to survive and pass their genes (and frightened amygdala) on to future generations. On the other hand, early humans who were too calm to worry about, say, whether a lion was nearby or whether a river looked like it would flood their dwelling were less likely to survive and pass on their genes. Through natural selection, humans living today are the descendants of people whose amygdalas produced very effective fear responses.

Having a protective, fear-producing amygdala is nearly universal among humans. It isn't surprising, therefore, that anxiety disorders are the most common mental disorder people experience, affecting approximately forty million adults in the United States (Kessler et al. 2005). Given that the daily dangers in our lives have been reduced tremendously since prehistoric times, you may wonder why so many people are experiencing anxiety-based problems. Unfortunately, the amygdala is still operating on the lessons it learned in prehistoric times. It still considers us to be potential prey for other animals or humans. It assumes that the best response to danger is running, fighting, or freezing, and it prepares the body to initiate these responses whether they're appropriate or not. But these fear responses don't fit the twenty-first-century situations that most of us live in, and they don't help us in the way they once did. For instance, people seem to be predisposed to fear snakes, spiders, and heights rather than cars, guns, and electrical outlets, even though the latter can be more deadly than the former. In addition, it also seems that some people's brains are more susceptible to this fear response, whether due to genetics or living through traumatic experiences.

The Anatomy of Anxiety

Neuroscience involves the study of the development, structure, and function of the nervous system, including the brain. In order to explain the neuroscience of anxiety, we need to provide you with a concise description of the anatomy of the brain, especially of the cortex and the amygdala. Having a grasp of how these important regions of the brain operate and the ways in which they relate to one another will help you understand what happens when the cortex or amygdala overreacts and creates anxiety. This basic knowledge of neuroscience will provide you with insight into how you can rewire your brain to resist anxiety.

The Cortex Pathway

We'll start with the cortex pathway because when people talk about the brain, they usually picture the wrinkled, gray outer layer of the brain known as the cerebral cortex. The cortex is the source of many of the human race's most impressive abilities. But, as we'll explain, these abilities also result in the cortex being capable of creating a great deal of anxiety.

The Cerebral Cortex

In humans, the cortex is larger and has more developed abilities than those of other animals. It's divided into two halves: the left hemisphere and the right hemisphere. Furthermore, it's divided into different sections, called *lobes*, that have different functions, such as processing vision, hearing, and other sensory information and putting it together to allow you to perceive the world. The cortex is the perceiving and thinking part of the brain—the part you're using to read and understand this book.

In addition to providing sights, sounds, and other perceptions, the cortex also attaches meaning and memories to those perceptions. So you don't just see an old man and hear his voice; rather, you recognize him as your grandfather and understand the specific meaning of the sounds he's making. And beyond providing you with the ability to understand and interpret situations, the cortex allows you to use logic and reasoning, produce language, use your imagination, and plan ways of responding to situations.

The cortex can also contribute to changing your responses to threatening situations, which is key when dealing with anxiety. The cortex is capable of evaluating the usefulness of various responses to the dangers you face. Thanks to the influence of your cortex, you can decide not to physically fight your boss if you feel you're in danger of being fired, or choose not to run away when you hear exploding fireworks. In fact, by reading this book, you're doing the very same thing: actively using your cortex to find different ways to cope with anxiety.

The cortex pathway to anxiety begins with your sense organs. Your eyes, ears, nose, taste buds, and even your skin are all sources of information about the world. All of your knowledge of the world has come through your sense organs and been interpreted by different parts of your cortex. When information comes in through your sense organs, it's directed to the *thalamus*, which is like the Grand Central Station of the brain (see figure 1). The thalamus is a central relay station that sends signals from your eyes, ears, and so on to the cortex. When information comes into the thalamus, it's sent out to the various lobes to be processed and interpreted. Then the information travels to other parts of the brain, including the frontal lobes (behind the forehead), where the information is put together so that you can perceive and understand the world.

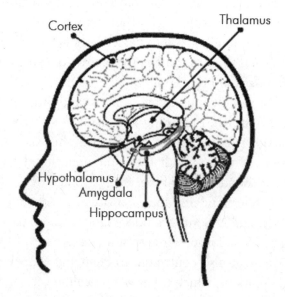

Figure 1. The human brain.

The Frontal Lobes

The *frontal lobes* are one of the most important parts of the cortex to understand. Located directly behind the forehead and eyes, they're the largest set of lobes in the human brain, and they're much larger than the frontal lobes of most other animals. The frontal lobes receive information from all of the other lobes and put it together to allow us to respond to an integrated experience of the world. The frontal lobes are said to have *executive functions*, meaning that they are where the supervision of many brain processes occurs. The frontal lobes help us anticipate the results of situations, plan our actions, initiate responses, and use feedback from the world to stop or change our behaviors. Unfortunately, these impressive capacities also lay the groundwork for anxiety to develop.

The cortex pathway is often a source of anxiety because the frontal lobes anticipate and interpret situations, and anticipation and interpretations often lead to anxiety. For example, anticipation can lead to another common cortex-based process that creates anxiety: worry. Because of our highly developed frontal lobes, humans have the ability to predict future events and imagine their consequences—unlike our pets, who seem to sleep peacefully without anticipating tomorrow's problems. Worry is an outgrowth of anticipation of negative outcomes in a situation. It's a cortex-based process that creates thoughts and images that provoke a great deal of fear and anxiety.

Some people have a cortex that's masterful at worrying, taking any situation and imagining dozens of negative outcomes. In fact, some of the most creative people are also sometimes the most anxious because their creativity gives them the ability to dwell on extremely frightening thoughts and images.

A common imagined worry among parents who have teenagers who are late for curfew (and what teenagers are not?) is imagining their kids injured in an accident, bleeding and unable to call for help. This image is terrifying—and completely unnecessary to envision—but some people seem to end up repeatedly anticipating these kinds of negative events. If your pattern of worrying is serious enough that it interferes with your daily life, you may be diagnosed with generalized anxiety disorder.

Another kind of anxiety disorder, obsessive-compulsive disorder, can occur when the frontal lobes create *obsessive thoughts*—cognitions or doubts that won't go away, to the point that people spend hours each day focused on them. Obsessions can sometimes lead a person to create elaborate rituals that must be carried out to reduce anxiety. Consider Jennifer, who thought obsessively about all of the germs in her house and spent hours washing her hands and cleaning certain areas of her home. Then, after she finished, she'd start over because she had doubts that led her to think that she might have touched something that contaminated everything that she'd cleaned. These kinds of obsessive thoughts may be due to a dysfunction in the *cingulate cortex*, an area in the frontal lobes just behind the eyes (Zurowski et al. 2012).

In summary, when we speak of the cortex pathway to anxiety, we're generally focused on interpretations, images, and worries that the cortex creates, or on anticipatory thoughts that create anxiety when no danger is present. As mentioned, when therapists assist people with modifying their thoughts to reduce worry, they're focused on the cortex pathway. Such cognitive approaches can be very effective at reducing cortex-initiated anxiety. However, as you now know, another neural pathway is also involved in the creation of anxiety, even when anxiety begins in the cortex.

The Amygdala Pathway

The second pathway involves the amygdala. Although the cortex pathway to anxiety may be more familiar or understandable because we are often aware of the thoughts it produces, the amygdala initiates the physical experience of anxiety. Its strategic location and connections throughout the brain enable it to control the release of hormones and activate areas of the brain that create the physical symptoms of anxiety. In this way, the amygdala exerts powerful and immediate effects on the body, and these are crucial to understand.

The Amygdala

The amygdala is located near the center of the brain (see figure 1). As previously stated, the brain actually has two amygdalas, one in the left hemisphere and one in the right; but it's customary to refer to the amygdala as singular, so we'll continue this practice. The position of your right amygdala can be estimated by pointing your left index finger at your right eye and your right index finger into your right ear canal. The point of intersection of the lines from your two fingers is about where your right amygdala is located. Because the amygdala is an almond-shaped structure, it gets its unusual name from the Greek word for almond.

The amygdala is the source of many of our emotional reactions, both positive and negative. When someone violates your personal space or gets in your face, it's the amygdala that produces the rage you feel. On the other hand, when you meet someone who looks like your grandmother and you experience a warm feeling of affection for this lady you don't even know, that's also the amygdala, in this case accessing a pleasant emotional memory. The amygdala both forms and recalls emotional memories; and if you understand this, your emotional reactions will probably make much more sense to you.

The Lateral Nucleus

The amygdala is divided into several sections, but we'll focus mainly on two that play essential roles in creating emotional responses, including fear and anxiety. The *lateral nucleus* is the part of the amygdala that receives incoming messages from the senses. It constantly scans your experiences and is at the ready to respond to any indication of danger. Like a built-in alarm system, its job is to identify any threat you see, hear, smell, or feel and then send a danger signal. It gets its information directly from the thalamus. In fact, it receives information *before* the cortex does, and this is important to keep in mind.

The reason the lateral nucleus gets information so fast is because the amygdala pathway is the more direct route from our senses. The amygdala is wired to respond quickly enough to save your life. Its rapid response is possible because of a shortcut in brain wiring that allows information to get to the lateral nucleus of the amygdala directly (Armony et al. 1995). When our eyes, ears, nose, or fingertips receive information, the information travels from these sense organs to the thalamus, and the thalamus sends this information directly to the amygdala. At the same time, the thalamus also sends the information to the appropriate areas of the cortex for higher-level processing. However, the amygdala receives information before the information can be *processed* by the various lobes in the cortex. This means the lateral nucleus of the amygdala can react to protect you from danger before your cortex even knows what the danger is.

See figure 2 for a simplified illustration of the pathways that allow the amygdala to react before the cortex can.

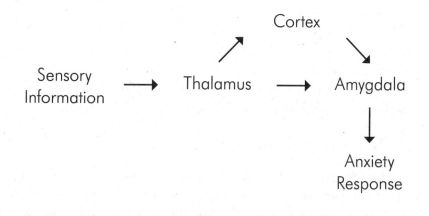

Figure 2. The two neural pathways to anxiety.

You can see the two pathways to anxiety in this illustration. Information goes directly from the thalamus to the amygdala, allowing the amygdala to react before you have time to use your cortex to think. While this may seem odd, if you consider your own experiences, you can probably recall some times when this has occurred. Have you ever been in a situation in which you reacted instinctively before you had time to know what you were reacting to?

Consider Melinda, a ten-year-old girl who was looking for camping equipment in the basement of her home. She walked through a doorway and jumped back in fear. Her reaction was triggered by a coat hanging on a coatrack. Her amygdala responded to the shape of the coat, which could have been an intruder, and caused her to jump out of reach of the "intruder" before she even realized what she'd seen. As an evolution-based safety measure, the amygdala is wired to react before the cortex can.

The detail-focused cortex takes more time to process information from the thalamus. In Melinda's case, the visual information needs to be sent to the occipital lobe at the back of the head, and from there it's sent to the frontal lobes, where the information is integrated and informed choices arise. That's why Melinda jumped back immediately but recovered in a moment and resumed looking for the camping equipment: it took a moment for her cortex to provide the information that the dark shape was a completely harmless coat. (You'll find a step-by-step explanation, including a downloadable figure, that illustrates the two pathways at work in Melinda's behavior at http://www.newharbinger.com/31137; see the back of the book for information on how to access it.)

The Central Nucleus

The amygdala can accomplish its quick response because of the special properties of another nucleus within it: the *central nucleus*. This small but powerful cluster of neurons has connections with a number of highly influential structures in the brain, including the hypothalamus and the brain stem. This circuit can signal the sympathetic nervous system to activate the release of hormones into the bloodstream, increase respiration, and activate muscles—all in a fraction of a second.

The close connection of the central nucleus to elements of the *sympathetic nervous system* (SNS) provides the amygdala with a great deal of influence over the body. The SNS is made up of neurons in the spinal cord that connect with nearly every organ system in the body, which allows the SNS to influence dozens of responses, from pupil dilation to heart rate. The role of the SNS is to create the fight-or-flight response, an effect that is balanced by the influences of the *parasympathetic nervous system* (PNS), which allows us to "rest and digest."

During fear-provoking situations, the lateral nucleus sends messages to the central nucleus to activate the SNS. At the same time, the central nucleus also activates the *hypothalamus*. (See figure 1 for the location of the hypothalamus.) The hypothalamus controls the release of cortisol and adrenaline, hormones that prepare the body for immediate action. These hormones are released from the adrenal glands, located atop your kidneys. *Cortisol* increases blood sugar levels, giving you the energy you need to use your muscles. *Adrenaline* (also called epinephrine) gives you an energetic feeling that heightens your senses, increases your heart rate and breathing, and can even keep you from experiencing pain. All of these responses come from the amygdala pathway.

Clearly, the amygdala wields a lot of power when it comes to initiating split-second physical reactions. In part, this is because the amygdala is strategically located in a central area of the brain, with immediate access to information from the senses and an advantageous position to influence parts of the brain that can change essential bodily functions very quickly. Being aware of how the amygdala functions is a crucial piece of the anxiety puzzle.

A Matter of Timing

As you can see, one clear distinction between the amygdala and the cortex is that they operate on different timetables. The amygdala can cause you to act on information sooner than your cortex can process it, orchestrating a bodily response before the cortex has even finished organizing the information for you to perceive it. While this is beneficial in some situations, the fact that we have little control over the amygdala's rapid responses means that we *experience* our fear and anxiety responses, rather than consciously controlling them.

The quick reaction that results from the amygdala pathway is typically called the *fight-or-flight response*. You're probably familiar with this phenomenon, which prepares the body to react quickly in a dangerous situation. Most of us have experienced this response and can recall times we felt an adrenaline rush and reacted in an unthinking, immediate way to protect ourselves from a threat. How many people have been saved on the freeway by lightning-quick, instinctive reactions arising in the amygdala? The central nucleus of the amygdala is where the fight-or-flight response is initiated.

Being aware of these rapid responses initiated by the amygdala pathway can help you understand and cope with the physical experience of anxiety, including the most extreme anxiety reaction: a panic attack. People who have panic disorder and suffer from panic attacks find it useful to recognize that many aspects of a panic attack are related to the amygdala's activation of the fight-or-flight response. Pounding heart, trembling, stomach distress, and hyperventilation are all related to the amygdala's attempts to prepare the body for action. These symptoms often cause people to think they might be having a stroke or heart attack or are "going crazy." When people understand that the roots of a panic attack often lie in the amygdala's attempts to prepare the body to respond to an emergency, they're less likely to be troubled by these concerns (Wilson 2009).

The reactions of fight or flight are the most familiar fear responses, but the amygdala can also produce another response to fear that's less recognized: freezing, or becoming very still. In fact, we prefer the term *fight, flight, or freeze response* because so many people say they feel paralyzed when under extreme stress. As strange as it seems, for our ancestors the reaction of freezing may have been as helpful as fighting or fleeing in certain situations. Like a rabbit that remains motionless as you walk your dog past her nest, those who freeze sometimes find an advantage in remaining still when threatened.

When you're experiencing the fight, flight, or freeze response, the amygdala is in the driver's seat and you're a passenger. That's why, in emergency situations, you often feel as though you're observing yourself responding rather than consciously controlling your response. There's a reason why we don't feel in control in these moments, or in control of our anxiety: the amygdala isn't just faster—it also has the neurological capability to *override* other brain processes (LeDoux 1996). There are many connections from the amygdala to the cortex, allowing the amygdala to strongly influence the cortex's responding on a variety of levels, while fewer connections travel from the cortex to the amygdala (LeDoux and Schiller 2009). Therefore, it's literally true that you can't think when the amygdala takes control. The thinking processes of the cortex are superseded and you're under the influence of the amygdala.

Although you may question the usefulness of this arrangement, in some situations it's crucial. Would it be wise for your brain to wait for the cortex to analyze the make, model, and color of a car crossing the center line toward you and consider details such as the facial expression of the driver before reacting? Clearly, the ability of the amygdala to override the cortex can literally save your life. In fact, it probably already has.

Being aware of the amygdala's ability to take over is crucial for anyone who's struggling with anxiety. It's a reminder that the brain is hardwired to allow the amygdala to seize control in times of danger. And because of this wiring, it's difficult to directly use reason-based thought processes arising in the higher levels of the cortex to control amygdala-based anxiety. You may have already recognized that your anxiety often doesn't make sense to your cortex, and that your cortex can't just reason it away.

In addition, the amygdala can also influence the cortex by causing the release of chemicals that influence the entire brain, including the cortex (LeDoux and Schiller 2009). These chemicals can literally change the way you think. Therefore, strategies for coping with amygdala-based anxiety are essential, even though cortex-focused approaches are more commonly offered in treatment. In part 2 of this book, you'll learn techniques for controlling amygdala-based anxiety responses.

Brain Circuitry

Based on what you've read thus far, you now know which parts of the brain are involved in different types of anxiety. You know that the cortex pathway produces worries, obsessions, and interpretations that create anxiety, and you know that the amygdala initiates bodily reactions that make up the fight, flight, or freeze response. Many people find some comfort in simply knowing where various symptoms are coming from, that their reactions make sense, and that they aren't going crazy.

Now that you understand the parts of the brain that are involved in creating anxiety, you're probably interested in how you can change the way these parts of the brain respond. In order to do so, you need to make changes in the brain's circuitry. The brain is made up of billions of connected cells that form circuits that hold your memories, produce your feelings, and initiate all of your actions. These cells are called *neurons*, or nerve cells, and they're the basic building blocks of the brain. They're the reason that your brain has neuroplasticity: the ability to change itself and its responses. On the basis of your experiences, the neurons in your brain are capable of changing their structures and patterns of responding. Understanding how neurons function will help you learn strategies that will allow you to rewire the circuits in your brain that create anxiety. It will also help you understand the effects of antianxiety medications on the brain.

Neurons

Neurons are composed of three basic parts (illustrated in figure 3). The *cell body* contains the machinery of the cell, including the genetic material that directs the building of the cell. Coming out of the cell body are *dendrites*, which look like branches of a tree. Dendrites are an essential part of the communication system between neurons. They reach out to other neurons to receive messages, which travel between neurons by means of a chemical process. Dendrites receive messages from the *axons* of other neurons. The axons don't touch the dendrites; rather, they send their messages by releasing chemicals called *neurotransmitters* into the space between the axon and the dendrite. Examples of neurotransmitters include adrenaline, dopamine, and serotonin.

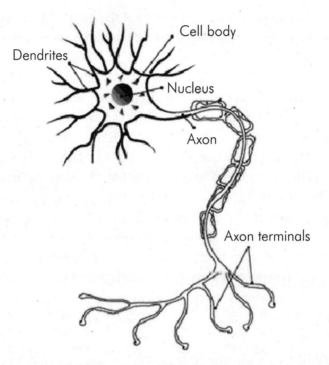

Figure 3. The anatomy of a neuron.

The space between an axon and dendrite is called a *synapse* (illustrated in figure 4). In this tiny space, communication between neurons occurs. At the end of the axon, called the *axon terminal*, tiny sacs hold neurotransmitters in preparation for sending chemical messages. Some neurotransmitters excite the next neuron, and others inhibit or quiet it.

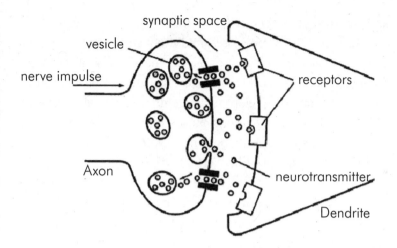

Figure 4. A synapse between two neurons.

Neurotransmitters are called chemical messengers because when they cross the synaptic space, it's as if they're taking a message to the next neuron. Neurotransmitters connect to *receptor sites* on the dendrites of the next neuron and have an effect similar to that of putting a key in a lock. We won't go into the specifics, but suffice it to say that when a neurotransmitter connects with the receptor site, it can cause the neuron to react by firing. *Firing* is when a positive charge travels from the receiving dendrites of a neuron, through the cell body, and all the way to axons at the other end. This causes the axon to release neurotransmitters from its axon terminals, transmitting the chemical message to yet another neuron, passing the message on.

Neurons operate on the basis of chemical messages between neurons and electrical charges within neurons. Every sensation you experience, from the sight of these words on the page to the sounds of birds singing in your yard, is processed in your brain by neurons. The sensations you experience, such as light waves that enter your eyes or vibrations in the air that impact your eardrums, get translated into electrical signals within neurons, and these signals are then communicated to other neurons via neurotransmitters. By means of these communication processes, the brain builds circuits of neurons that work together to store memories, create emotional reactions, initiate thought processes, and produce actions.

When scientists discovered that the messages sent between neurons were based on neurotransmitters sent from one neuron to the next, they began to develop medications that could target this process. Many of the medications most commonly used to treat anxiety, such as Lexapro (escitalopram), Zoloft (sertraline), Effexor (venlafaxine), and Cymbalta (duloxetine), were designed to increase the amount of neurotransmitters available in the synapse as a way of affecting circuits in certain areas of the brain. (The specific ways these medications affect neurons and how they influence anxiety are explained in "Medications and Your Anxious Brain," a bonus chapter of this book available for download at http://www.newharbinger .com/31137.)

Circuitry: Connections Between Neurons

Why do you need to know how neurons operate? If you want to rewire the brain, it helps to understand the brain's circuitry and its basis in connections formed between neurons. A Canadian psychologist named Donald Hebb (1949) proposed a theory of how neurons create circuitry that's turned out to be very useful in explaining the process. His idea has since been distilled into this simple statement by neuroscientist Carla Shatz: "Neurons that fire together wire together" (Doidge 2007, 63). This statement offers clear insight into how you can change the wiring in your brain.

Basically, for neurons to build connections between themselves, one neuron needs to be firing at the same time another neuron is firing. When neurons fire together, a connection between them is strengthened, and eventually a pattern of circuitry develops in which activation of one neuron causes the other to activate as well. More neurons can be connected with these neurons in a similar way, and if they fire together, soon a whole set of connected neurons is created. Changing neural circuitry involves changing the activation patterns in the brain so that new connections develop between neurons and new circuits form. Changes in the brain, or learning, can occur as a result of neurons establishing new connections and circuitry.

Although our brains are programmed from birth to develop and organize themselves, they're amazingly flexible and exquisitely responsive to the particular experiences of each individual. As neuroscientist Joseph LeDoux (2002, 3) explains it, "People don't come preassembled, but are glued together by life." The circuitry in your brain is shaped by the specific experiences you've had, and it can be changed as a result of your continuing experiences. For example, connections between particular neurons are strengthened when you use them. Some of us continue to use our memories of multiplication tables to calculate math equations, and those connections remain as strong as when we were in school. But some of us rely on calculators, so we don't regularly use the brain circuits storing the multiplication tables, and our memory of these tables weakens.

The specific circuitry in your brain develops based on the experiences you have. Perhaps your brain came to associate horses with stables, grandfathers with cigars, the smell of popcorn with baseball, and so on. Although two people may share similar associations, each of us has uniquely formed brain circuits based on our own experiences. While one person may have circuitry that associates cows with cheese and Wisconsin, another person may have circuitry that associates cows with barns and milking machines.

Neurons make new connections and build new circuits in a variety of ways. Circuitry can be activated by certain deliberate thoughts, like those you have when you're asked to remember your grandmother. Circuitry can be reorganized by changing your behavior, such as learning a new golf swing. Performing behaviors, like playing the piano or serving a volleyball, can cause new circuits to develop, and even imagining performing these behaviors can cause changes in circuitry. The brain remains flexible and capable of making changes throughout life.

If you want to change the anxiety you experience, you need to change the neural connections that lead to anxiety responses. Some of these connections are stored in the brain's circuitry in the form of memories, and memories are formed in both the cortex and the amygdala.

Emotional Memories Formed by the Amygdala

Emotional memories are made by the lateral nucleus in the amygdala through the process of association, which we'll discuss in chapter 2. These emotional memories come from experiences that your cortex may or may not remember. This is because the memory system in the cortex is completely separate from that in the amygdala. In fact, evidence suggests that amygdala-based memory is longer lasting than cortex-based memory (LeDoux 2000). In other words, the cortex is much more likely than the amygdala to forget information or have trouble retrieving it.

The existence of different memory systems explains why you can experience anxiety in a situation without any conscious memory (or understanding) of why the situation produces anxiety. Just because your amygdala has an emotional memory of an event doesn't mean that your cortex remembers the same event. And if your cortex doesn't remember the event, you'll have difficulty remembering it because we humans rely on our cortex memory. This means we sometimes have emotional reactions that puzzle us, especially when it comes to anxiety. So you may not understand why crossing bridges is anxiety provoking, why you avoid sitting with your back to the door in a restaurant, or why the smell of tomato plants makes you tense.

The amygdala is capable of reacting on the basis of its own emotional memories and doesn't need cortex-based memories. Research tracing the pathways in the brain that give rise to emotional responses has shown that emotional learning can occur without involvement of the cortex (LeDoux 1996). Here's an example that will help illustrate this (from Claparede 1951).

A woman was hospitalized as a result of Korsakoff's syndrome, a memory disorder often associated with chronic alcoholism. Her cortex couldn't form memories of her experiences, so she couldn't identify her doctor or the hospital, despite the fact that she'd been in the same hospital for years. She didn't know the name of the nurse who had cared for her for months, and she couldn't remember details of a story told to her only minutes before. But at the same time, her amygdala demonstrated the ability to create emotional memories without the aid of her cortex.

One day her physician performed a little experiment (one that wouldn't be ethical by today's standards). When he reached out to shake her hand, he stuck her hand with a pin he'd hidden in his palm. The next day, when the woman saw the doctor extend his hand, she quickly withdrew her hand in fear. When asked why she refused to shake his hand, she couldn't offer an explanation. In addition, she reported that she had no memory of seeing the doctor before. She had no cortex-based memory of an event that would cause her to fear the doctor; but her amygdala had created an emotional memory, and her fear was the evidence of it.

Discovering the Source of Amygdala-Based Memories

If you fear a specific object or situation, you may be able to recall an experience in which your amygdala learned that fear. Alternatively, it may be difficult to uncover how an amygdala-based fear developed, since your cortex isn't able to retrieve a memory related to that situation, even though the amygdala clearly does. The fact that the cortex can be left out of the loop is why people are often confused by their emotional responses.

Here's an example to illustrate what this confusion is like: Lily recognized that she had social phobia when she learned about the symptoms of social phobia on an anxiety website. She knew that she felt uncomfortable in groups of people and that it was difficult for her to attend family gatherings, like Thanksgiving dinner or a sister-in-law's baby shower. When her therapist told her that this anxiety was probably due to her amygdala, she had no idea why her amygdala had developed this emotional response. But after her therapist asked her to identify the specific characteristics of the gatherings that provoked anxiety, Lily said that being in a circle of people, even cheerful family members, was very distressing. She recognized that a circle of people was terrifying to her, especially when they all could look at her at once.

When the therapist asked Lily if she could think of an experience that could have taught her amygdala that a circle of people was dangerous, Lily recalled an event in second grade when she was in a circle of children reading aloud from their books. When it was her turn to read, she had difficulty, and the teacher treated her in a way that made her feel humiliated. The cortex-based memory of this experience finally came back to Lily, and she understood why her amygdala had created an emotional memory to try to protect her. Because of that memory, her amygdala responded to a circle of people as though they posed a danger.

Realizing that your amygdala stores emotional memories that your cortex has no knowledge of can help you better understand some of your emotional reactions. Sometimes the cortex has a complete lack of understanding of the origins or purposes of the emotional reactions created by the amygdala. But you can learn more about these processes, so in the next chapter we'll help you and your cortex become more knowledgeable about the workings of the amygdala.

Summary

Two pathways can create anxiety. One pathway travels the detail-focused circuitry of the cortex and eventually sends information to the amygdala, which produces an anxiety response. The other pathway runs directly from the thalamus to the amygdala. Each pathway can cause the amygdala to create anxiety, but each is also constructed of circuitry, and certain aspects of that circuitry can be modified. If you understand how the circuitry works, you can rewire your anxious brain so that you experience less anxiety.

CHAPTER 2

The Root of Anxiety: Understanding the Amygdala

Don't be fooled by the small size of your amygdala. Even though the largest and most developed portion of the human brain, the cortex, contributes to anxiety in many ways, the amygdala plays the most influential role because, as you learned in chapter 1, it's involved in both the cortex pathway and the amygdala pathway to anxiety. Like the conductor of an orchestra, the amygdala controls many different reactions in both your brain and your body. In addition to relying on preprogrammed responses, it's also exquisitely sensitive to what happens to you and responds to your specific experiences.

In this chapter, you'll learn about the amygdala's special "language" and its impact on your life. In evolutionary terms, the amygdala is an extremely ancient structure, and the human amygdala is quite similar to the amygdalas found in all other animals. Because the human amygdala is so similar to that in rats, dogs, and even fish, researchers have been able to study its functioning in depth and have learned a great deal about how it creates fear and anxiety.

When you're born, your amygdala has preprogrammed responses that are ready to be put into action. But this ancient structure isn't fixed; the amygdala is constantly learning and changing based on your day-to-day experiences. In fact, once you understand what we call the "language of the amygdala," you'll have more control over your anxiety responses because you'll know how to influence the part of the brain that's at the very root of fear.

The Amygdala as Protector

To understand amygdala-based anxiety, it's useful to think of the amygdala as your protector. Natural selection has given humans a fear-producing amygdala that has protection as a central goal. As you go about your day, the amygdala is vigilant for anything that might indicate potential harm. While the goal of protection is good, the amygdala can overreact, creating a fear response in situations that aren't really dangerous.

Consider Fran, who's about to give a speech. Her heart begins to pound and she starts to hyperventilate as soon as she stands in front of the group with everyone staring at her. What is her amygdala trying to protect her from? It seems as if it sees the situation of standing in front of an audience as dangerous.

Fran is not alone in experiencing this type of response. Studies have shown that fear of public speaking is the most commonly reported fear, surpassing fear of flying, fear of spiders, fear of heights, and fear of tightly enclosed spaces (Dwyer and Davidson 2012). What could account for this common response? Because the amygdala tries to prevent us from being prey to a predator, evolutionary scientists have suggested that we may be prone to interpreting eyes watching us as a potentially dangerous situation (Ohman 2007). Others have suggested that the risk of rejection by a group of observers comes from an ancient fear of being rejected by one's clan (Croston 2012), which once meant being left on your own to fend for yourself and face roaming predators—a likely death sentence. In any case, it appears that the human amygdala reacts to protect us from being in the vulnerable situation of being observed by potentially hostile animals, including other humans.

Fran may not be aware of the evolutionary roots of her reaction and the amygdala's role in it. Her cortex may be telling her that she's afraid of being criticized, humiliated, or making a mistake, while her amygdala is operating from a more prehistoric perspective. The truth is, the cortex often comes up with reasons for our behaviors, which may or may not be accurate explanations. However, the concern here isn't the accuracy of the cortex but its effects. The more Fran dwells on cortex-generated explanations for her amygdala-based anxiety, such as *You're worried that your boss will never be satisfied with this presentation*, the more cortex-based anxiety she'll create, adding to her problem. Looking in the cortex for the causes of amygdala-based anxiety is like looking in your refrigerator to understand why your car won't start. You aren't looking in the right place!

Instead, Fran needs to focus on her amygdala's perspective. She needs to see that her amygdala is trying to protect her. Instead of using her cortex to seek explanations for her anxiety, she needs to use her cortex to apply her knowledge of the language of the amygdala. First, she needs to recognize that her pounding heart and increased rate of breathing, which would help her if she needed to run or fight, don't indicate that she's truly in danger. These responses are part of the amygdala's reaction, and they aren't helpful in the context of public speaking. Fran needs to understand that this isn't a dangerous situation and that her amygdala is setting off an alarm unnecessarily. Even if the speech Fran is about to give is very important, perhaps for her career, it's unlikely that this is the life-or-death situation that her amygdala seems to be preparing her for.

This underscores the importance of being aware of the amygdala's role as protector. This is crucial in understanding and controlling your own anxiety responses. In many cases, the amygdala's assumption that you need to be protected from danger is incorrect. Fortunately, you can remedy this by retraining your amygdala, and by not giving the amygdala more fuel for the fire by assuming that a fearful or anxious emotional reaction is a definite indication of danger. The protective amygdala reaction is often misguided, and you don't want to let your cortex strengthen the reaction.

Lastly, it's important to recognize that simply trying to use your cortex to convince yourself that the situation isn't truly dangerous won't always shut off the amygdala's response. A more effective approach is using deep breathing techniques and strategies that retrain the amygdala, and we'll outline this approach in part 2 of the book, Taking Control of Your Amygdala-Based Anxiety.

How the Amygdala Decides What's Dangerous

The human amygdala seems predisposed to respond to some stimuli as if they're dangerous (Ohman and Mineka 2001). Fears of snakes, insects, animals, heights, angry facial expressions, and contamination seem to be biologically wired into the amygdala, because humans learn them with very little prompting. For example, few children have a car phobia, but many are afraid of insects. Although cars pose a much greater danger to children than insects, the fear of insects seems to be hardwired into the amygdala such that children develop this fear very easily. This is undoubtedly the result of thousands of years of evolution in which a fear of insects contributed in some way to survival. However, even fears that are programmed into the amygdala can be changed. If they couldn't, it's unlikely that so many of us would live with sharp-toothed animals like cats or dogs and treat them as part of our families.

On the other hand, many objects or situations aren't naturally feared by the amygdala. Instead, the amygdala learns to fear them as a result of life experiences. The amygdala is constantly learning based on experience, and after certain negative experiences, it creates brain circuits that cause people to fear a previously unfeared object. For example, a child doesn't naturally fear a flame and must be warned not to touch it. But after a child is burned by, say, a birthday candle, the child's amygdala learns to fear the sight of flames. In addition, the amygdala quickly adds various flaming objects to its list of dangerous things to avoid, so the child may also fear lighters, sparklers, and campfires. The amygdala retains lasting memories that identify the object and similar objects as dangerous. This is a very powerful and adaptive ability, because it allows for the creation of specialized neural circuitry that helps people avoid the specific dangers that occur in their lives. This has kept the amygdala useful and virtually unchanged for millions of years.

When we explain the two pathways to anxiety to people, they often ask if they could have inherited a sensitive amygdala. Genetics can definitely influence the amygdala and therefore your typical emotional reactions. For example, children who have a smaller left amygdala tend to have more anxiety difficulties than other children (Milham et al. 2005). The good news is, every amygdala is capable of learning and changing, and in later chapters you'll learn how to train your amygdala to respond differently.

Emotional Memories

As discussed in chapter 1, the amygdala forms memories, but not in the way people typically think about memories. On the basis of your experiences, your amygdala creates emotional memories—both positive and negative—that you don't necessarily have an awareness of. Positive emotional memories, such as the association of the smell of perfume with feelings of love for your partner, usually don't cause much difficulty. Therefore, we'll focus on negative emotional memories, especially those that result in fear and anxiety, because these memories can cause a great deal of amygdala-based anxiety.

As noted in chapter 1, the lateral nucleus of your amygdala creates emotional memories based on your experiences, and these memories can lead you to respond to certain objects or situations as if they're dangerous. Because of these memories, you're conscious of a feeling of discomfort, fear, or dread. However, you don't realize that this feeling is due to an emotional memory because the memory isn't stored as an image or verbal information. It isn't like an old photograph or movie in your mind, as cortex-based memories can be. Instead, you experience an amygdala-based memory *directly*, as an emotional state. You simply begin feeling a specific emotion. If this feeling is anxiety, it's easy to assume that having a fearful or anxious feeling is an accurate reflection of the dangerousness of a situation—if you don't understand the language of the amygdala.

Consider Sam, who was in a car accident that severely injured his girlfriend, who was driving. To this day, when he rides in the passenger seat of a car, Sam has anxiety—an urgent feeling of danger that seems to result from the current situation in his environment.

Sam doesn't experience a memory of the accident or reflect on the accident every time he experiences this amygdala-based anxiety. However, anytime he approaches the passenger seat, he has a strong feeling that he must avoid the situation, and he becomes extremely uncomfortable and almost panicky whenever he tries to ride with another driver. If he tries to put his feelings into words, he says it feels like something bad is going to happen if he rides in the passenger seat. He's more comfortable driving himself, and for years he's avoided being a passenger. Because his deeply felt emotional reaction is so real and persistent, he doesn't consider whether he should question it. He would never describe it as a memory formed by his amygdala, and he doesn't expect himself to change it or even realize that he can.

Exercise: Getting Familiar with the Effects of Amygdala-Based Memories

You may wonder what an amygdala-based memory feels like. Read through the list of experiences below and consider whether any of them are similar to something you've felt. Check any that apply to you:

_____ I notice my heart pounding strongly or my heart rate increasing in certain situations.

_____ I avoid certain experiences, situations, or locations without consciously intending to do so.

_____ I keep watching or checking on certain things even when I don't really need to.

_____ I can't relax or let my guard down in a specific location or type of location.

_____ Seemingly insignificant events can make me worry.

_____ I can end up in a complete panic very quickly.

_____ In certain situations, I feel very angry to the point of wanting to physically fight, but I know that doesn't make sense.

_____ *I feel a strong urge to escape from certain situations.*

_____ *I feel overwhelmed and can't think clearly in certain settings.*

_____ *Under certain circumstances, I feel paralyzed and can't get myself to do anything.*

_____ *In stressful situations, I can't breathe at a normal rate.*

_____ *I develop a lot of muscle tension in certain situations.*

All of the statements above reflect possible effects of memories formed in the lateral nucleus of the amygdala. If you've felt some of these reactions, you were probably experiencing the influence of an amygdala-based memory. Your amygdala may have stored these memories in an attempt to protect you from potential danger. When these memories are activated, it's normal not to understand the reaction you're having or not to feel in control of your responses. Furthermore, you may have come up with erroneous explanations for these reactions, which are based on your cortex's desire to understand what's happening.

The Fight, Flight, or Freeze Response

As mentioned earlier, the amygdala's central location in the brain places it in an advantageous position to influence other parts of the brain that can change essential bodily functions in a fraction of a second. When danger is detected, the amygdala can affect a number of highly influential structures in the brain, including the brain stem arousal systems, the hypothalamus, the hippocampus, and the nucleus accumbens. These direct connections allow the amygdala to instantly activate motor (movement) systems, energize the sympathetic nervous system, increase levels of neurotransmitters, and release hormones like adrenaline and cortisol into the bloodstream. This activation creates a cascade of changes in the body: heart rate increases, pupils dilate, blood flow is shifted away from the digestive tract to the extremities, muscles tense, and the body is energized and primed for action. In response to these physiological changes, you may feel trembling, a pounding heart, and stomach and bowel distress.

All of these changes are part of the fight, flight, or freeze response, and as noted previously, the central nucleus is the portion of the amygdala where the fight, flight, or freeze response is initiated. When this reaction is needed, we consider it a lifesaving event. But if the central nucleus overreacts, it can set off a full-blown panic attack when no logical reason for fear exists.

Once it initiates a panic attack, the central nucleus of the amygdala is in control and the cortex has very little influence. Some people respond very aggressively when they panic, some flee the situation, and others are immobilized. If you're having a panic attack and people try to provide you with logical reasons why you shouldn't be panicking, they're essentially talking to a cortex that's turned off. Strategies that directly target the amygdala, such as physical activity or deep breathing, will be more effective, and we'll teach you these strategies in chapters 6 and 9.

Being aware of the amygdala's ability to take charge is essential information for anyone who struggles with anxiety. It serves as a reminder that everyone's brain is hardwired to allow the amygdala to seize control in times of danger. Countless lives (of humans and other animals) have been saved by the amygdala's ability to quickly commandeer bodily reactions in dangerous situations. Examples include slamming on the brakes in traffic, ducking when a foul ball is headed your way, or leaving the room when the veins in your boss's neck are bulging. All of these situations are instances when your amygdala is attempting to save you from perceived danger. But as we mentioned, sometimes the well-intentioned amygdala is itself the problem.

The Language of the Amygdala

At this point, you've learned a great deal about how the amygdala creates anxiety. You understand that one of the main functions of the amygdala is to protect you. You also know that it can identify certain objects or situations as dangerous, usually because of a learning experience. You learned that the amygdala creates memories that you may not be aware of, but which you experience as emotions. Finally, you know that the amygdala has an immediate response system that can take over both your brain and your body when it feels that you're in danger. This raises the question of how we can exercise control over the amygdala. To do so, we need to communicate new information to this small but powerful part of the brain, and the best way achieve this is by using the amygdala's own language.

We use the word "language" here to describe the method of communication between the amygdala and the outside world. This particular language is not one of words or thoughts, but of emotions. When it comes to anxiety, the language of the amygdala has a fairly narrow focus on danger and safety. It's based on experience, and it's a language of quick action and response. When you understand the specifics of this language, your experiences with amygdala-based anxiety will make more sense, and you'll also be able to communicate new information to your amygdala in order to train it to respond differently.

As discussed in chapter 1, a central law underlying the neural circuits in the brain is "neurons that fire together wire together" (Doidge 2007, 63). The amygdala's language is based on creating connections between neurons. When it comes to amygdala-based anxiety, connections between neurons are made when sensory information about an object or situation is being processed by neurons in the lateral nucleus of the amygdala at the same time that something threatening happens to excite the amygdala. In any threatening situation, the amygdala is working to identify any sight, sound, or other sensory information associated with danger. Therefore, *association* is an essential part of the language of the amygdala.

Psychologists have known about association-based learning, typically called classical conditioning, for over a century, but only in the past couple of decades have they recognized that some types of this learning occur in the amygdala. In this book, we make use of many findings from neuroscientist Joseph LeDoux (1996) and his team, who are researching the neurological basis of amygdala-based anxiety. The amygdala scans the sensory aspects of your life and responds in very specific ways when sensory information is associ-ated with positive or negative events occurring at the same time. When sensations, objects, or situations have been associated with a negative event, memories are stored by the lateral nucleus in circuits that are wired to produce a negative emotion.

Emotional Learning in the Lateral Nucleus of the Amygdala

Imagine a person being confronted by a dog. The sights and sounds of the dog are processed through the thalamus and relayed directly to the lateral nucleus of the amygdala, which doesn't automati-cally create a change in neural circuitry that will cause anxiety. Neurons in the lateral nucleus change in such a way that fear is learned only if sensory information about the dog is being processed in the lateral nucleus *immediately before or at the same time as* the occurrence of a negative experience, such as being threatened or bitten by the dog. Thus, if the dog behaves in a friendly or neutral manner, the lateral nucleus won't create a negative emotional memory about the dog.

However, when a painful or negative experience such as a dog bite occurs, the neurons transmitting the sensory information about the bite create strong emotional excitation in the lateral nucleus. If this excitation is occurring at about the same time that the lateral nucleus is receiving sensory information about the dog, the lateral nucleus changes neural circuitry to respond negatively to dogs or similar animals in the future. In studies on rats, scientists have actually been able to observe that connections form in the amygdala when such pairings are experienced (Quirk, Repa, and LeDoux 1995).

An object or situation itself need not be harmful or threatening for fear or anxiety to be associated with it. Any object, even a teddy bear, can come to cause anxiety through association-based learning. For an association to develop, all that's required is that the object be experienced at about the same time that some arousing or threaten-ing event is activating the lateral nucleus. Remember, neurons become connected when they're firing at the same time.

The association-based language of the amygdala is what creates many of the emotional reactions you experience; amygdala-based anxiety is only one example. In the case of anxiety, the lateral nucleus connects sensory information from a situation with the emotion of fear. After this connection has been created, you'll feel anxious whenever the amygdala recognizes similar sensory informa- tion. The sights, sounds, or smells associated with the negative event become capable of activating the amygdala's alarm system. The term *trigger* refers to anything—an event, object, sound, smell, and so on—that activates the amygdala's alarm system as a result of association-based learning. In the example above, dogs become a trigger for anxiety. Triggers are an important aspect of the language of the amygdala.

It may seem surprising that *any* object can become a trigger if it's processed when the amygdala is in an activated state. But amygdala- based anxiety is due to associations, not logic, so triggers need not make logical sense. Here's an example that illustrates how associa- tion, not cause and effect, governs amygdala-based anxiety: Josefina was presenting a teddy bear to her grandson, who was running happily toward her. Then he suddenly fell and split his lip open on the driveway. Now he experiences amygdala-based anxiety whenever he sees a teddy bear. Because the perfectly harmless teddy bear was associated with the pain of the injury, the teddy bear became a trigger, leading to a fear of teddy bears.

The amygdala's reaction may range from relatively weak to very strong depending on the experience. For example, you might have a mild dislike of a certain type of food that was associated with a nega- tive experience, such as egg salad that you ate during a stressful family picnic. On the other hand, if you once ate pancakes when you had an illness that later caused you to throw up, you might find that, even years later, just the smell of pancakes makes you nauseous.

Before you get the idea that you might be better off without your amygdala, remember that its role is to protect you. In addition, it produces positive emotions due to association-based learning. For example, if your special someone gives you a necklace as a gift, you'll experience feelings of warmth and love for your partner. Later, when you see the necklace, the association formed between the necklace and the emotion of love will make you experience warm, affectionate feelings again. Had the necklace not been paired with a loved one, it would simply be another piece of jewelry. Many positive emotional reactions are produced by the amygdala, so you wouldn't want to get rid of it.

In fact, if two people have had different experiences, they can have completely different reactions to the same object, thanks to the language of the amygdala. One of the authors (Catherine) has affectionate feelings for daddy longlegs because she frequently encountered them while picking her favorite red raspberries in her grandmother's garden. She has been known to gently pick up daddy longlegs and take them out of her home, much to the horror of her coauthor (Elizabeth), whose amygdala reacts to daddy longlegs as though they're frightful.

Exercise: Identifying Amygdala Emotions in Your Life

Can you think of harmless situations or objects that elicit amygdala-based anxiety as a result of the association-based language of your amygdala? Have you ever been puzzled by your reaction to something or someone you had no good reason to fear or dislike? Also consider whether you've ever experienced unexpected positive emotions in response to someone or something. These emotional responses could be a reflection of the language of the amygdala. On a separate piece of paper, list examples of both positive and negative reactions. Remember, the items you list for either category need not make logical sense. For example, you may have a negative emotional reaction to the scent of lilacs and a positive emotional reaction to lightning storms.

The Amygdala's Reactions Aren't Logical

As you can see, amygdala-based emotions aren't rational. They're based on associations, not logic. Consider Beth, who was sexually assaulted while a specific Rolling Stones song was playing. After the assault, whenever Beth heard the song she felt intense anxiety. Obviously, the Rolling Stones song had nothing to do with the sexual assault; it was just a coincidence that it was playing when the assault occurred. Nonetheless, Beth's amygdala responded to the association between the song and the assault, an extremely negative event. In this way, the amygdala transforms a neutral object or situation into something that creates an emotional reaction. To be more accurate, the object itself isn't transformed; rather, it's processed in a new or different way by the amygdala.

People *experience* the connection the amygdala makes between an object and fear, but they may not recognize or understand the connection. They may feel a strong emotional reaction to an object without realizing that a neural connection has been made or understanding why the emotional reaction is occurring. This lack of awareness is completely normal and extends to all sorts of neural functions. For example, you don't have to be consciously aware of the neural circuits that allow you to read this book, to sit upright, or to breathe. Thank goodness! That kind of awareness would be exhausting.

However, for people who suffer with anxiety, having an understanding of the amygdala's significant role in creating fear associations is helpful. It allows you to stop looking for logical explanations and start learning to use the language of the amygdala. We'll use Don, a Vietnam veteran with post-traumatic stress disorder (PTSD) as an example of how having a grasp of the language of the amygdala can be helpful. Don used to experience panic attacks but then didn't have one for many years. Suddenly, he started having a panic attack each morning for no apparent reason. When encouraged to investigate the situation, Don realized that his panic was closely associated with showering. After a few days of observing his anxiety build as he showered, Don realized that his wife had switched to the same brand of soap that he'd used in Vietnam. The smell of the soap was activating an amygdala response and creating panic attacks. In the language of the amygdala, the soap was a trigger associated with the war.

Recognizing that the soap was the reason for his panic attacks was a relief for Don. Knowing the language of the amygdala gave him a new understanding that helped him see that he wasn't going crazy and that his PTSD wasn't starting to take over his life again— something he was very concerned about. In Don's case, understanding the language of the amygdala was helpful, even though it didn't end his anxiety. He still felt anxious whenever he smelled the soap, despite knowing that the soap wasn't dangerous; however, he could put an end to his morning panic attacks by switching to a different brand of soap.

For Don, avoiding that brand of soap had no costs. But sometimes the trigger is something more difficult, or impossible, to avoid. Consider a plumber who has a fear of spiders (which have a way of hiding under sinks), or an office manager who works on the twentieth floor and has panic attacks in elevators. In these cases, reducing or eliminating the fear or panic attacks requires retraining the amygdala. We'll explain how to do this in part 2 of the book; for now, simply know that there are ways to change your emotional circuitry. This can be an enormous source of hope.

It may be that you aren't sure where a certain emotional reaction came from. Fortunately, it isn't necessary to know the original cause of amygdala-based anxiety in order to change the emotional circuitry. As you'll see in chapter 7, once you recognize that a specific trigger is associated with an anxiety response, you can take steps to change the circuitry associated with that trigger, even if you don't know the original cause of the emotional memory.

Learning from Experience

Many people believe that the symptoms of anxiety disorders, such as panic, worry, and avoidance of certain objects or situations, should be alleviated by rational argument. Well-meaning family members and friends, and sometimes even people struggling with anxiety, often think logic and reason should change the way the anxious person reacts. But, of course, the amygdala isn't logical. For example, if a young boy fears dogs after being bit by one, you won't get very far by saying, "Don't worry about my dog Buddy. He's never bitten anyone. He's all bark and no bite." Once you have a grasp of the language of the amygdala, it's clear why logic-based interventions miss the mark. As you'll see later in the book, many cortex-based anxiety symptoms do respond to logical arguments, but when it comes to amygdala-based anxiety, there's only one sure way for the amygdala to learn: experience.

The amygdala's reliance on experience for learning explains why hours of talk therapy or working through numerous self-help books may not help with anxiety: they may not be targeting the amygdala. If you want the amygdala to change its response to an object (for example, a mouse) or a situation (such as a noisy crowd), the amygdala needs experience with the object or situation for new learning to occur. Experience is most effective when the person interacts directly with the object or situation, although observing another person has also been shown to affect the amygdala (Olsson, Nearing, and Phelps 2007). You can reason with the amygdala for hours, but if you're trying to change amygdala-based anxiety, that tactic won't be as effective as a few minutes of direct experience will be.

So, to change your amygdala's fear response to, say, a mouse, you must be in the presence of a mouse in order activate the memory circuits related to mice. Only then can new connections be made. Because the amygdala learns on the basis of associations or pairings, it must *experience* a change in these pairings for the circuitry to change. Not surprisingly, when your mouse-memory circuits are activated, you're going to feel some anxiety.

Unfortunately, people typically try to avoid such experiences, and this avoidance prevents the amygdala from forming new connections. Returning to the example of the mouse, you may even try to avoid *thinking* about mice, because just the thought of a mouse can cause the amygdala to react, initiating an anxiety response. The amygdala tends to preserve learned emotional reactions by avoiding any exposure to the trigger, which decreases the likelihood of changing that emotional circuitry. Being the ultimate survivalist, the amygdala is purposely cautious, and its default setting is to organize responses that decrease your exposure to triggers. But again, amygdala-based anxiety responses won't change if the amygdala is successful in avoiding triggers.

When you come to terms with the idea that you need to activate the amygdala's circuits to generate new associations, you've learned an important lesson. We like the pithy phrase "activate to generate" as shorthand for this requirement, which is perhaps the most challenging lesson in the language of the amygdala. It's challenging because it involves accepting the experience of anxiety as necessary for new learning to occur. By engaging in experiences that activate the amygdala's memory of a specific object or situation, you communicate to the amygdala in its own language and put it in the best situation for new circuits to form and new learning to occur.

Summary

In this chapter, you've learned how the amygdala creates anxiety as a result of the associations it experiences. You've learned that one of the main functions of the amygdala is to protect you, and that the amygdala creates memories that you may not be aware of but experience instead as emotional reactions. The amygdala has an immediate response system that can take over both your brain and your body when it feels you're in danger. But the amygdala can learn from its experiences, and you can use the amygdala's own language of associations to make new connections. In chapters 7 and 8, you'll learn how to rewire the amygdala so that it responds in a calm way. If you've suffered with the mysteries of amygdala-based anxiety for years, this will provide an amazing feeling of empowerment.

CHAPTER 3

How the Cortex
Creates Anxiety

Although the amygdala pathway is very powerful in its ability to activate a variety of physical reactions instantly, anxiety can also have its origins in the cortex pathway. The cortex operates in a completely different way than the amygdala, but its responses and circuitry can prompt the amygdala to produce anxiety. Through this process, the cortex can create unnecessary anxiety and also worsen anxiety that originates in the amygdala. Once you understand how your cortex initiates or contributes to anxiety, you can see the possibilities for either interrupting or modifying cortex reactions to reduce your anxiety.

Origins of Anxiety in the Cortex

The cortex can initiate anxiety in two general ways. The first involves how the cortex processes sensory information, such as sights and sounds. As discussed, the thalamus directs sensory information to the cortex, as well as to the amygdala. As the cortex processes this information, it can interpret perfectly safe sensations as threatening. It then sends a message along to the amygdala that can produce anxiety. In this case, the cortex turns a rather neutral experience that wouldn't naturally activate the amygdala into a threat, causing the amygdala to react by creating an anxiety response.

Here's an example: A high school senior who had applied to several colleges looked at the mail and saw an envelope from one of the colleges he'd applied to. Imagining that it contained a rejection letter, he had a few very anxious moments before opening the envelope. As it turned out, he'd been admitted and had even been awarded a scholarship. Nevertheless, his cortex initiated an anxiety response by interpreting the sight of the envelope in a way that created distressing thoughts, and these thoughts activated his amygdala. This type of cortex-based anxiety depends on the cortex's interpretation of the sensory information it receives.

The second general way the cortex can initiate an anxiety response occurs without the involvement of any specific external sensations. For example, when worries or distressing thoughts are produced in the cortex, this can activate the amygdala to produce an anxiety response even though the person hasn't seen, heard, or felt anything that's dangerous in any way. An example would be parents of an infant who leave their little boy with a babysitter to go out for dinner and suddenly begin to have concerns about their child's safety. Even though the boy is perfectly safe, the parents imagine that he's in distress or being neglected by the sitter. Thoughts and images like these can activate the amygdala even though there's no sensory information indicating that the child is in danger.

Cognitive Fusion

Before we examine these two general ways in which the cortex creates anxiety, we want to address a process that can occur in both: *cognitive fusion*, or believing in the absolute truth of mere thoughts. It's one of the biggest problems created in the cortex, which can produce a rigid belief that thoughts and emotions should be treated as though they reflect an ultimate reality that can't be questioned. Both the high school senior and the worried parents in the examples above may have fallen victim to cognitive fusion by taking their negative thoughts and images too seriously.

Confusing a thought with reality is a very seductive process due to the cortex's tendency to believe it possesses the real meaning of every thought, emotion, or physical sensation. Actually, the cortex is surprisingly prone to misinterpretations and errors. It's common to have erroneous, unrealistic, or illogical thoughts or to experience emotions that don't make much sense. In reality, you need not take every thought or emotion you have seriously. You can allow many thoughts and emotions to simply pass without undue attention or analysis. In chapter 11, we'll discuss cognitive fusion in detail, help you assess whether you're prone to cognitive fusion, and provide strategies to help you defuse from thoughts.

Anxiety That Arises Independently from Sensory Information

Now we'll take a closer look at the different ways the cortex can initiate anxiety. First we'll consider the type of anxiety that begins as thoughts or images produced by the cortex, without any information from your senses. There are actually two subcategories of this process—thought-based and imagery-based—and each typically arises in a different hemisphere of the cortex, with thought-based anxiety coming from the left hemisphere and imagery-based anxiety coming from the right. That said, these two types of cortex-induced anxiety aren't mutually exclusive. In fact, they often occur together.

Left Hemisphere–Based Anxiety

Distressing thoughts are more likely to come from the left side of the cortex, which is the dominant hemisphere for language in most people. Logical reasoning, which is produced in the left hemisphere, underlies both worry and verbal rumination (Engels et al. 2007). Worry is the process of envisioning negative outcomes for a situation. *Rumination* is a style of thinking that involves repetitively mulling over problems, relationships, or possible conflicts. In rumination, there's an intense focus on the details and possible causes or effects of situations (Nolen-Hoeksema 2000). Although people may believe thinking processes like worry or rumination will lead to a solution, what actually happens is a strengthening of the circuitry in the cortex that produces anxiety. In addition, rumination has been shown to lead to depression (Nolen-Hoeksema 2000).

Whatever you devote a great deal of time to thinking about or think about in great detail is more likely to be strengthened in your cortex. The circuits in the brain operate on the principle of "survival of the busiest" (Schwartz and Begley 2003, 17), and whatever circuitry you use repetitively is likely to be very easily activated in the future. This means that instead of leading to solutions, the processes of worry and rumination create deep grooves in your thinking processes that make you tend to focus on those concerns in your left hemisphere. Sometimes people get lost in repeatedly analyzing situations, creating an experience called *anxious apprehension* (Engels et al. 2007). As these persistent, worrisome thoughts are rehearsed repeatedly in the mind, they become increasingly difficult to dismiss. This type of thinking is especially common among people with generalized anxiety disorder and obsessive-compulsive disorder.

Right Hemisphere–Based Anxiety

The human ability to imagine situations in detail comes from the right hemisphere of the cortex, which approaches the world differently than the analytical, verbal left hemisphere. The right hemisphere is nonverbal and processes things in more holistic, integrated ways. It helps us see patterns, recognize faces, and identify and express emotions. It also provides us with visual images, imagination, daydreams, and intuition. Because of these capacities, it can contribute to anxiety based on imagination and visualization.

When you visually imagine something frightening, you use your right hemisphere to do so. When you hear the critical tone of accusations in your imagination, your right hemisphere is involved. If you're particularly good at using your imagination, you can expect your amygdala to respond. The amygdala can become highly activated when the right hemisphere creates frightening images.

Research suggests that the right hemisphere is strongly connected to anxiety symptoms (Keller et al. 2000). In fact, it's more strongly associated than the left hemisphere with the kind of anxiety in which a person feels strong arousal and intense fear (Engels et al. 2007). For example, people with panic disorder are more likely to have right hemisphere–based anxiety (Nitschke, Heller, and Miller 2000). So when you're feeling strong, arousing anxiety, as opposed to apprehensive or worry-based anxiety, the right side of your cortex is more likely to be activated. *Vigilance*, a general state of alertness in which the whole environment is scanned for indications of danger, is also based in the right hemisphere (Warm, Matthews, and Parasuraman 2009).

Anxiety That Arises from the Cortex's Interpretations of Sensory Information

Now we'll turn to the other type of cortex-based anxiety described at the beginning of the chapter: anxiety that arises from the cortex's *interpretations* of otherwise neutral sensory information. Sometimes you may be in a situation that's perfectly safe, but your cortex responds to sensory information as if it's dangerous or upsetting. Information coming from your senses via the thalamus is given meaning by the way that the circuits in the cortex process and interpret that information. Let's revisit the example of the high school student who thought he was being rejected by a college but was actually being offered a scholarship. His cortex had interpreted an envelope as a source of distressing news and turned it into a very frightening object.

The frontal lobes of the human cortex have a well-developed capacity to contemplate future events and imagine their consequences. This is often quite helpful, with the cortex producing interpretations that allow us to respond well to a variety of situations.

However, problems begin when the cortex repeatedly reacts in ways that produce anxiety. Whether due to certain learning experiences, specific physiological processes, or, most frequently, a combination of both, the circuitry in the cortex can respond in ways that promote worry, pessimism, and other negative interpretive processes. (We'll discuss this in greater detail in part 3 of the book.)

If your cortex interprets a perfectly safe situation as threatening, you'll feel anxiety. Consider Damon, who's walking his dog in his neighborhood. He sees a fire truck heading in the direction of his house with its lights on and sirens blaring and interprets this to mean that his house is on fire. As a result, he starts to feel tremendous anxiety. The cause of his anxiety is his cortex's interpretation of the meaning of the fire truck, not the fire truck itself. (Figure 5 illustrates this process.)

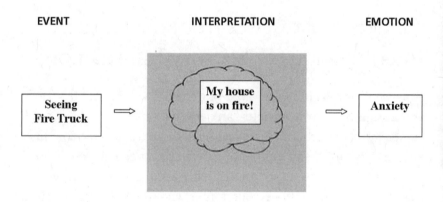

Figure 5. How the cortex's interpretations can create anxiety.

Figure 5 makes it clear that the thoughts produced in Damon's cortex, not the actual event of seeing the fire truck, are what created his anxiety. In fact, from Damon's location it's impossible to have information that confirms that his house, or anyone's house, is on fire, so there's no reason that seeing this fire truck should produce anxiety. It may be reasonable for his cortex to come to the conclusion that there could be a fire, but other explanations exist, such as an accident or medical emergency that has nothing to do with a fire or Damon's house. But instead of considering these options, Damon imagines that his house is on fire. As a result, his left hemisphere gets to work considering the ways a fire could have started, thinking, *I may have left the stove on* or *The wiring is so old. Maybe a short started a fire.* Meanwhile, his right hemisphere is creating images of his kitchen engulfed in flames. His amygdala is likely to react to these kinds of thoughts and images, and in response, Damon may rush home in a panic, even though there's no actual threat to his home. His interpretation is the source of his anxiety.

Anticipation: The Gift of the Human Cortex

Because the human cortex has the ability to predict future events and imagine their consequences, we experience anticipation, which is both a blessing and a curse. *Anticipation*, which refers to expectations about what will occur, is based on the cortex's ability to begin preparing for a future event by considering or visualizing it. It occurs primarily in the prefrontal cortex (which lies behind the forehead) on the left, more verbal side. The left prefrontal cortex is the part of the brain where we plan and execute actions, so it's not surprising that anticipation arises here, as it's about getting ready to act in some way. We can anticipate in positive ways and feel excited and eager about an upcoming event. However, we can also anticipate in negative ways, expecting and imagining negative or even dangerous events. This can lead to a great deal of distress.

The anticipation of negative situations creates threatening thoughts and images that can significantly increase anxiety. In fact, the experience of anticipation is often more distressing than the anticipated event itself! In many cases, the thoughts and images people have about an upcoming situation, such as a potential confrontation, an exam, or a task that must be completed, are much worse than the actual situation turns out to be.

As you can see, the cortex's ability to use language, produce images, and imagine the future allows it to initiate an anxiety response in the amygdala even when no reason for anxiety exists. People usually find it easier to recognize the cortex's role in creating anxiety than the amygdala's role. This is because we're more able to observe and understand the language of thoughts and images produced in the cortex. Some parts of the cortex are more directly under our control than the amygdala is, and we're more able to interrupt and change cortex-created thoughts and images. That said, we don't mean to suggest that controlling the cortex is easy. Your cortex has established certain patterns of responding, and once it has developed these habits, it can be challenging to interrupt and change them. But they can be changed, and we'll explain how you can accomplish this in part 3 of the book.

The Final Step in the Cortex Pathway to Anxiety: The Amygdala

Discussion of the cortex pathway isn't complete until we address the role of the final component in the pathway: the amygdala. On its own, the cortex can't produce an anxiety response; the amygdala and other parts of the brain are needed to accomplish that. In fact, people without a functioning amygdala, whether due to stroke, illness, or injury, don't experience fear in the way most people do.

Consider the case of a woman whose amygdalas were both destroyed by a rare condition, Urbach-Wiethe disease (Feinstein et al. 2011). Her story offers a glimpse of what life is like without the amygdala's fear response. She can be exposed to spiders or snakes or watch terrifying scenes from horror movies without experiencing fear. Even more remarkably, in the course of her daily life she was held up at gunpoint and also was almost killed during an assault but experienced no fear in either situation. In fact, she's been the victim of a variety of crimes, probably because she lacks the caution that would arise from a functioning amygdala. Her experiences illustrate that the amygdala is the source of the fear response. No matter what thoughts, images, or expectations originate in the cortex, many of the emotional and physiological aspects of anxiety result only when the cortex activates the amygdala.

The amygdala responds to information passed on from the cortex. In fact, the amygdala may respond to what we *imagine* in much the same way that it responds to what's actually happening. Information based on thoughts or images of potential danger travels the same pathways as information associated with actual perceptions and interpretations. As discussed earlier, the amygdala almost instantly processes information it receives directly from the senses via the thalamus. After a delay during which the cortex processes and interprets the information, the amygdala also receives information from the cortex. Neuroscientists don't yet know exactly how the amygdala distinguishes whether the information it receives from the cortex is valid or based on an overactive imagination.

Let's look at two examples of how the amygdala might respond to thoughts or images created in the cortex to examine how the amygdala's reliance on the cortex can be either beneficial or problematic. In the first example, Charlotte is at home one evening when she hears the familiar sound of someone coming in the back door. She hears this noise every night when her husband comes home, so her amygdala doesn't respond to the sound as a signal of danger. But Charlotte knows *in her cortex* that her husband is away on a fishing trip and that no one should be coming in the back door at this time. Her cortex produces thoughts of danger and an image of a stranger entering her home. These thoughts and images in Charlotte's cortex influence her amygdala to initiate the fight, flight, or freeze response. Charlotte's heart starts pounding and she stops what she's doing. She becomes hypervigilant and focuses on getting herself to safety. If there is an intruder, these reactions could save her life.

Charlotte's amygdala isn't responding to the sound of the door. It's responding to Charlotte's *thoughts* that there may be a stranger in the house. Responding to information from the cortex allows the amygdala to guard against dangers it doesn't recognize. The amygdala relies on the cortex to provide it with additional information.

But sometimes the amygdala's reliance on the cortex leads to unnecessary anxiety, as in the next example.

In this scenario, Charlotte is once again alone at home while her husband is away. She doesn't hear anything unusual, but she feels uneasy when she goes to bed. As she lies in bed listening to the quiet night, she imagines that someone is breaking into the house. She imagines an intruder walking around inside the house carrying a weapon, and her amygdala responds to these images in her cortex. Even though there's no direct evidence that she's in any danger, her amygdala still responds to the activity in her cortex by initiating the fight, flight, or freeze response. Suddenly, Charlotte feels a terrible sense of dread. Her breathing becomes shallow and she feels she should hide or seek help, even though she *realizes* that there's no strong evidence of danger.

Charlotte's amygdala is responding to the thoughts and images in her cortex as if they reflect actual danger, and it creates a very real fear response. As you can see from these two examples, what you think about and focus on in the cortex can definitely affect your level of anxiety. From the perspective of the amygdala, thoughts or images in the cortex may call for a response, even if the amygdala itself doesn't detect danger from the sensory information it received more directly. In reacting to the cortex's information, the amygdala may initiate the fight, flight, or freeze response. And once the amygdala gets involved, you begin to experience the physical sensations associated with anxiety.

Fortunately, a number of techniques can be used to interrupt and change cortex-based thoughts and images that may activate the amygdala. With practice, you can rewire your cortex to be less likely to activate your amygdala. The first step is to recognize when the cortex is producing thoughts or images that may lead to anxiety. When you become aware of these thoughts and their anxiety-inducing effects, you can begin to recognize the thoughts, identify when they occur, and take steps to change them.

Summary

At this point, you're familiar with a variety of ways that anxiety can be initiated in the cortex. You've seen that the amygdala can be activated by thoughts from the left hemisphere or images from the right hemisphere. You've also been informed of the dangers of cognitive fusion and learned ways that the cortex's interpretations and anticipation can lead the amygdala to create anxiety. In part 3 of the book, we'll examine specific cortex-based interpretations and reactions that can lead to anxiety and discuss strategies that will help you change the thoughts and images your cortex produces. But first, in the next chapter, we'll help you consider various aspects of your anxiety and pinpoint whether it originates primarily in your cortex or amygdala. This is a key step in determining how to rewire your brain to control your anxiety. Once you identify the starting point of your anxiety, you can apply the right techniques to effectively address the problem.

CHAPTER 4

Identifying the Basis of Your Anxiety: Amygdala, Cortex, or Both?

Anxiety is a complex response that, in most cases, involves a variety of areas of the brain. While the amygdala and cortex both play a role, it's helpful to know where your own anxiety begins. This determines which strategies will be most helpful in reducing it. In this chapter, we'll help you assess whether your anxiety is cortex-based, amygdala-based, or both. You'll also learn more about how your anxious thoughts and reactions affect you and your life.

Where Does Your Anxiety Begin?

Based on earlier chapters, you now know that even though the amygdala is the neurological source of the anxiety response, creating the physical sensations of anxiety and often overriding cortex-based thought processes, anxiety doesn't always begin in the amygdala. It can also begin in the cortex, with thoughts and mental images activating the amygdala. If you become anxious when you see a growling dog and begin to hyperventilate, that would be amygdala-initiated anxiety. If you're pacing nervously as you anticipate an important phone call, that would be cortex-initiated anxiety. Understanding where and how your anxiety begins will allow you to take the most effective approach to interrupting the process.

It's important to remember that when anxiety begins in the amygdala, cortex-based interventions, such as logic and reasoning, don't always help reduce anxiety. Amygdala-based anxiety can often be identified by certain characteristics; for example, it seems to come from out of the blue, it creates strong physiological responses, and it seems out of proportion to the situation. When anxiety starts in the amygdala, you need to use the language of the amygdala to modify it. Amygdala-initiated anxiety is most effectively reduced by the interventions in part 2 of this book, "Taking Control of Your Amygdala-Based Anxiety."

If, on the other hand, you know that your anxiety began in the cortex, the more effective approach is to change your thoughts and images to decrease the resulting amygdala activation. You'll learn about how you can accomplish this in part 3 of the book, "Taking Control of Your Cortex-Based Anxiety." Decreasing the number of times your cortex causes your amygdala to become activated will cut down on your overall anxiety.

The rest of this chapter consists of informal assessments that will help you evaluate and describe your typical anxiety responses to assist you in determining where your anxiety originates. Please note that these aren't professional assessments; they're simply provided to help you explore your amygdala- and cortex-based tendencies.

Cortex-Based Anxiety

We'll start by addressing anxiety initiated by circuitry in the cortex. Certain types of activation in the cortex, often experienced as thoughts or images, can eventually cause the amygdala to activate the stress response, along with all of its unpleasant symptoms. The varieties of cortex-based activation are numerous, but they all have the same potential consequence: putting you at risk for experiencing anxiety. The following assessments will provide more insight into some of the most common ways the cortex pathway can initiate anxiety and will help you identify which ones you experience. Typically, people don't pay close attention to the specific thoughts and images occurring in their cortex, so it is essential that you become more watchful and aware of what's happening in your cortex at any given moment. By learning to recognize different types of anxiety-provoking cortex activities, you can modify them before they escalate into full-blown anxiety. We'll explain how to do so in part 3 of the book.

Exercise: Assessing Left Hemisphere–Based Anxiety

As explained in chapter 3, the left hemisphere of the cortex can produce a type of anxious apprehension that shows up as a tendency to worry about what will happen and search repetitively for solutions. With this type of anxiety, people tend to ruminate or focus intensely on a situation or feel the need to discuss a situation repeatedly.

Read through the examples below and check those that describe you:

_____ I rehearse potential problem situations in my mind, considering various ways things could go wrong and how I'll react.

_____ I often think about situations from the past and consider ways they could have gone better.

_____ I tend to get stuck in the process of considering different ways I could talk to someone about concerns or other topics.

_____ Sometimes I just can't turn off a stream of negative thinking, and it often prevents me from sleeping.

_____ I find it comforting to consider a problem from a number of different perspectives.

_____ I feel much better when I have a solution for a possible difficulty, just in case the situation arises.

_____ *I know I tend to dwell on difficulties, but it's just because I'm trying to find explanations for them.*

_____ *I have difficulty getting myself to stop thinking about things that make me anxious.*

If you checked several of the items above, you may be spending too much time focusing on distressing situations and bringing to mind thoughts that increase your level of anxiety. Although your left hemisphere may be looking for a solution, a strong focus on potential difficulties can activate the amygdala. You may be missing many opportunities for anxiety-free moments by thinking about problems that might never occur.

The left hemisphere provides us with some of our most complex and highly developed abilities, and we humans couldn't have created the technologically sophisticated world we live in without its contributions. But the worry and rumination it creates don't provide the solution to anxiety. In part 3 of the book, we'll take a closer look at various ways the left hemisphere contributes to anxiety. We'll help you identify specific kinds of thought processes that lead to anxiety, such as pessimism, worry, obsessions, perfectionism, catastrophizing, and guilt and shame, and explain how you can change these thought processes.

Exercise: Assessing Right Hemisphere–Based Anxiety

The right hemisphere of the cortex allows you to use your imagination to visualize events that aren't actually occurring. Imagining distressing situations can activate the amygdala. The right hemisphere's focus on nonverbal aspects of human interactions, such as facial expressions, tone of voice, or body language, may cause you to jump to conclusions about this information. For example, it's easy to make too much of a facial expression or a gesture and assume someone is angry or disappointed.

Read through the statements below and check any that you experience often:

_____ *I picture potential problem situations in my mind, imagining various ways things could go wrong and how others will react.*

_____ *I'm very attuned to the tone of people's voices.*

_____ *I can almost always imagine several scenarios that illustrate how a situation could turn out badly for me.*

_____ *I tend to imagine ways that people will criticize or reject me.*

_____ *I often imagine ways that I might embarrass myself.*

_____ *I sometimes see images of terrible events occurring.*

_____ *I rely on my intuition to know what others are feeling and thinking.*

_____ *I'm watchful of people's body language and pick up on subtle cues.*

If you checked many of the statements above, your anxiety may be increased by a tendency to imagine frightening scenarios or rely on intuitive interpretations of people's thoughts that may not be accurate. These right hemisphere–based processes can cause your amygdala to respond as if you're in a dangerous situation when no threat exists. A variety of strategies, including play, exercise, meditation, and imagery can be useful for increasing activation of the left hemisphere, producing positive emotions, and quieting the right hemisphere. We'll discuss these strategies in chapters 6, 9, 10, and 11.

Exercise: Identifying Anxiety That Arises from Interpretations

In chapter 3, we discussed how the interpretation of events, situations, and other people's responses can lead to anxiety. When this occurs, a person's cortex is creating unnecessary anxiety. The anxiety is being produced not by the situation, but by the way the cortex is interpreting the situation.

To determine if your cortex has a tendency to turn neutral situations into sources of anxiety, read through the list below and check any items that apply to you:

_____ *I have a tendency to expect the worst.*

_____ *I think I take people's comments too personally.*

_____ *I have trouble accepting the fact that I make mistakes, and I beat myself up when I do.*

_____ *I have a hard time saying no because I don't like to disappoint people.*

_____ *When I have a setback, I find it overwhelming and feel like giving up.*

_____ *When I have trouble finding something, I worry that I'll never find it.*

_____ *I tend to focus on any flaws in my appearance.*

_____ *When someone makes a suggestion, I can't help but consider it a criticism.*

If you checked many of the statements in the list above, the interpretations provided by your cortex are probably increasing your anxiety. Many people believe that certain situations are the cause of their anxiety, but anxiety always begins in the brain, not with the situation. Anxiety is a human emotion, produced by the human brain, and emotions are caused by the brain's reaction to situations, not the situations themselves. People have different reactions to the same event because of their differing interpretations. For example, seeing a wolf in the woods may terrify a camper but fascinate a zoologist. How your cortex interprets events can obviously have a strong impact on how much anxiety you experience. In chapters 10 and 11, you'll learn how to resist anxiety-producing interpretations.

Exercise: Assessing Your Anxiety Based in Anticipation

When you anticipate, you're using your cortex to think about or imagine future events. If those future events have the potential to be negative, anticipation can serve to increase anxiety. As with left hemisphere–based anxiety, this can lead to anxiety about things that might not ever occur. And even if the event does come to pass, you may start dwelling on it long before it occurs or you need to be concerned about it. So instead of experiencing the event just once, you experience it repeatedly before it ever occurs.

Here are some statements that reflect a tendency to anticipate. Read through the list and check any that apply to you:

_____ If I know a potential conflict is looming, I spend a lot of time considering it.

_____ I think about things that people might say that would upset me.

_____ I can almost always think of several ways that a situation could turn out badly for me.

_____ When I know that something might go wrong, it's constantly on my mind.

_____ I can be worried sick about something months before it occurs.

_____ If I'm going to have to perform or speak in front of a group, I can't stop thinking about it.

_____ If there's a potential for danger or illness, I feel like I need to consider it.

_____ I often waste time thinking of solutions for problems that never occur.

If you have a tendency to anticipate negative events, you're creating more anxiety in your life than is necessary. Keep in mind that, while everyone experiences difficult situations in life, there's no need to live through these events in the cortex when nothing negative is occurring. We'll cover strategies for modifying your thoughts in chapter 11.

Exercise: Assessing Your Anxiety Based in Obsessions

When people have obsessions (repetitive, uncontrollable thoughts or doubts), perhaps accompanied by compulsions (activities or rituals performed in an effort to reduce anxiety), these behaviors arise in the cortex and are fueled by the anxiety of the amygdala. Obsessions, which are very much a product of the frontal lobe of the cortex, have been linked to excessive activation of the circuitry in the orbitofrontal cortex, an area just behind the eyes (Zurowski et al. 2012).

Read through the following statements, which reflect both obsessions and compulsions, and check any that apply to you:

_____ *I devote a great deal of thought to keeping things in order or doing tasks correctly.*

_____ *I'm preoccupied with checking or arranging things until I believe they're right.*

_____ *I'm haunted by certain doubts that I can't escape.*

_____ *I have concerns about contamination and germs.*

_____ *I have some thoughts that I find unacceptable.*

_____ *I worry about acting on urges that come into my mind.*

_____ *I get stuck on a certain idea, doubt, or thought and can't get past it.*

_____ *I have routines that I need to complete in order for things to feel right.*

If you checked several items, consider whether you're spending a lot of your time focusing on thoughts or activities that keep you stuck in patterns that maintain your anxiety in the long run and rob you of precious time. Obsessive thoughts can occur without compulsive behaviors, but often compulsions form when a person finds that these behaviors provide temporary relief from anxiety. Unfortunately, even though the compulsions don't help in the long run, they can be maintained by the amygdala because of the temporary relief from anxiety that follows them. Therefore, coping with obsessions and compulsions usually requires an approach that targets the amygdala as well as the cortex. We'll discuss ways of dealing with cortex-based obsessions in part 3, and explain exposure methods that combat amygdala-fueled compulsions in chapter 8.

Amygdala-Based Anxiety

Now that you've identified cortex-based causes of your anxiety, we'll help you assess your tendency toward amygdala-initiated anxiety. As a reminder, anytime you feel anxiety or fear, the amygdala is involved. However, the following assessments will help you zero in on experiences where your anxiety response *originated* in the amygdala. Once you know the starting point, you can choose approaches that will best control your anxiety. If the circuitry in the amygdala itself is what initiated your anxiety, strategies that target the cortex will be futile. In part 2 of the book, we'll provide a number of techniques that are helpful for controlling amygdala-based anxiety, including relaxation strategies, exposure to feared objects or situations, engaging in physical activity, and improving your sleep patterns.

To determine whether the amygdala or the cortex initiated a specific anxiety response, you need to consider what was happening before you began to experience anxiety. If you were focusing on specific thoughts or images, that suggests your anxiety began in the cortex. If, on the other hand, you feel that a specific object, location, or situation immediately elicited an anxiety response, the amygdala is more likely to be the starting point.

Exercise: Assessing Your Experience of Unexplained Anxiety

When your anxiety seems unexplained or comes from out of the blue and you aren't able to find any good reason for it, your amygdala is probably the cause. You might honestly say, "I just don't know why I feel this way; it doesn't make sense," because none of your thoughts or current experiences justify the feeling. As we've noted, the amygdala often responds without your having any conscious awareness of what's happening, and the responses it creates are often puzzling.

Read through the following statements, which reflect unexplained anxiety, and check any that apply to you:

_____ *Sometimes my heart pounds for no reason.*

_____ *When I visit others, I frequently want to go home, even though things are going fine.*

_____ *I often don't feel in control of my emotional reactions.*

_____ *I can't explain why I react the way I do in many situations.*

_____ *I have sudden rushes of anxiety that seem to come from out of nowhere.*

_____ *I just don't feel comfortable going to certain places, but I don't have a good reason for feeling that way.*

_____ *I frequently feel panicky with no warning.*

_____ *I usually can't identify the triggers of my anxiety.*

As we've noted, you may not have access to the amygdala's memories. As a result, when your amygdala reacts you may have no clue what it's reacting to or why. The good news is, even when you don't understand why your amygdala is responding, you can choose from a wide variety of techniques to help calm your amygdala and rewire it.

Exercise: Assessing Your Experience of Rapid Physiological Responding

When the amygdala is the source of your anxiety, you're more likely to have noticeable physiological changes as one of the first signs of your anxiety. Before you have time to think or even fully process the situation, you may experience a pounding heart, sweating, and a dry mouth. Because the amygdala is strongly wired to energize the sympathetic nervous system, activate muscles, and release adrenaline into the bloodstream, having physiological symptoms as the first sign of anxiety is a good indicator that you're dealing with amygdala-based anxiety.

Read through the following statements, which reflect rapid physiological responding, and check any that apply to you:

_____ I find that my heart is racing even when there's no obvious reason.

_____ I can go from feeling calm to being in a complete panic in a matter of seconds.

_____ I suddenly can't get my breathing rhythm to feel right.

_____ Sometimes I feel dizzy or as though I might faint, and these feelings arise quickly.

_____ My stomach lurches and I feel nauseous right away.

_____ I become aware of my heart because I have pain or discomfort in my chest.

_____ I start sweating without exerting myself.

_____ I have no idea what comes over me. I just start trembling without warning.

If you checked many of these statements, which reflect strong and rapid physiological responding, your anxiety may originate in reactivity of the amygdala. When you experience such responses, you may assume that an actual threat is present. But your amygdala could be reacting to a trigger that isn't an accurate indicator of danger, so remember that a *feeling* of danger doesn't necessarily indicate the presence of a threat. You can use these physiological responses as an indication that you should use the strategies suggested in part 2 of this book.

Exercise: Assessing Your Experience of Unplanned Aggressive Feelings or Behavior

A tendency toward aggression is based on the fight element of the fight, flight, or freeze response. Whereas some people want to retreat and avoid conflicts or threatening situations, others tend to have aggressive responses. Suddenly feeling threatened can make them prone to anger and lashing out at others. This aggressive response, which has its roots in the protective nature of the amygdala, is especially charac teristic of people with post-traumatic stress disorder.

Read through the following statements, which reflect unplanned aggressive feelings or behavior, and check any that apply to you:

_____ *I explode unexpectedly in certain situations.*

_____ *I often need to do something physical to express my frustration.*

_____ *I strike out and later realize that my response was too strong.*

_____ *I snap at others with little warning.*

_____ *I feel that I'm capable of hurting someone when I'm under stress.*

_____ *I don't want to lash out at people, but I can't help it.*

_____ *Family members and friends know to be cautious around me.*

_____ *When I've been upset, I've broken or thrown objects.*

If you checked several of these statements, which reflect a tendency to show signs of anxious aggression, the amygdala-based interventions in part 2 of the book will be helpful. Your amygdala's attempts to activate an aggressive response can seem compelling, but you can exert control in how you direct your behavior. Regular physical exercise can help curb this kind of responding, and taking a brisk walk to get out of a threatening situation can help satisfy the drive to take immediate action.

Exercise: Assessing Your Experience of Inability to Think Clearly

When you find yourself not just anxious but also unable to concentrate or direct the focus of your attention, this is a strong indicator of amygdala-based anxiety. When the amygdala steps in, it overrides the attentional control of the cortex and takes charge. When you experience this amygdala-based control of your brain, you'll feel unable to control your thoughts. Remember, from an evolutionary standpoint the amygdala's ability to seize control when it detects danger helped our distant ancestors survive. Therefore, the amygdala has retained this capacity. Still, it's both disconcerting and frustrating to temporarily lose the ability to decide what to focus on or think about.

Read through the following statements, which reflect an inability to think clearly, and check any that apply to you:

_____ When I'm under pressure, my mind goes blank and I can't think.

_____ I know that when I'm anxious, I'm unable to focus on what I need to do.

_____ When I get nervous, sometimes I can't concentrate very well.

_____ When I'm being yelled at, I'm unable to come up with a response.

_____ When I feel panicky, it's often difficult for me to focus on what I need to do.

_____ Even when I try to calm down, it's hard for me to distract myself from how my body is feeling.

_____ When I'm scared, sometimes I draw a total blank about what I should do next.

_____ *During a test, I often can't remember what I've learned, even when I'm prepared.*

If you checked several of these statements, you may frequently find yourself in situations where you have an inability to think. The connections from the amygdala to the cortex can influence how attention is directed, and evidence suggests that people who experience high levels of anxiety often have weaker connections from the cortex to the amygdala (Kim et al. 2011). Cortex-based strategies for coping with anxiety are often not very useful when the amygdala is activated. Some of the strategies discussed in part 2 of the book, such as deep breathing or relaxation, will be helpful even when your thought processes are limited by activation of the amygdala.

Exercise: Assessing Your Experience of Extreme Responses

If your responses often seem over-the-top and out of proportion to the situation at hand, your amygdala is probably behind this pattern of extreme responding. It may be taking over and acting to protect you from a danger that it perceives, but which you'd recognize, in a calmer moment, as not requiring such a strong response. One of the most intense types of extreme response is a panic attack (discussed further in chapter 5), but there are others. In all cases, these extreme responses are caused by activation of the fight, flight, or freeze response when it isn't necessary. Remember, the amygdala's approach to situations is typically "better safe than sorry," and it's programmed to react swiftly and strongly—even when it isn't completely sure of the details involved in possible threats.

Read through the following statements, which reflect a pattern of extreme responses, and check any that apply to you:

_____ *At times, my anxiety is so strong that I'm afraid I'm going crazy.*

_____ *I get paralyzed by the level of anxiety I experience.*

_____ *Other people have told me they think I overreact.*

_____ *When something is out of place or disorganized, I can't tolerate it.*

_____ *At times, I've wondered whether I'm having a heart attack or stroke.*

_____ *Sometimes I just lose my temper and go into a rage.*

_____ *Little things, like an insect or dirty dishes, can send me into a complete panic.*

_____ *Sometimes things around me don't seem real, and I fear I'm losing my mind.*

If you checked several of these statements, you're probably suffering from excessive amygdala activation. As we noted earlier in the book, some amygdalas are more reactive than others, even quite early in life. Unfortunately, children with reactive amygdalas don't necessarily learn amygdala-based strategies for dealing with their anxiety, and the result is often entrenched patterns of overreacting or extreme avoidance. But as you've learned, it's never too late for the amygdala to learn to respond differently.

Summary

In the first half of this chapter, you assessed your tendency to experience cortex-based anxiety and determined whether specific thought processes are contributing to your anxiety. In the second half of the chapter, you assessed whether you're prone to experiences of amygdala-based anxiety: unexplained anxiety, rapid physiological responding, unplanned aggressive feelings or behavior, inability to think clearly, and extreme responses. Now that you have a better idea of where your anxiety originates—in your cortex, amygdala, or both—you're ready to look more closely at the nature of each type of anxiety and learn techniques that will help you minimize or control your specific anxiety responses.

PART 2

Taking Control of Your Amygdala-Based Anxiety

CHAPTER 5

The Stress Response and Panic Attacks

In this part of the book, we'll focus on the amygdala's role in producing an anxiety response and examine the influence of the amygdala pathway in more detail. As a reminder, the amygdala is always involved in creating an anxiety response, whether that response begins in the cortex or the amygdala. Therefore, understanding the amygdala will benefit anyone who has anxiety. Because feelings of anxiety result when the amygdala creates a stress response, we start by describing the stress response. Knowledge about this response and how it's controlled by the amygdala is essential to understanding the grip of fear or anxiety and how to free yourself from it.

In chapter 1, we noted that a certain area of the amygdala, the central nucleus, can initiate the fight, flight, or freeze response, causing an amazing number of changes in the body in an instant; and that such changes are beyond your control. We also explained in chapters 1 and 2 that when the central nucleus produces a strong fight, flight, or freeze response, your ability to use your cortex to think and respond is often limited. This is why it's essential to be able to recognize and understand the fight, flight, or freeze response *before* it occurs—and to learn ways to respond to it appropriately. Once you're experiencing it, your ability to use your cortex to cope with anxiety is diminished.

The Stress Response

The fight, flight, or freeze pattern of responding was first recognized by physiologist Walter Cannon (1929). Then, in the 1930s, endocrinologist Hans Selye recognized that animals and humans alike have a surprisingly similar reaction to a broad set of stressors. Our bodies usually respond in ways that are specific to given situations. For example, our pupils contract in bright light but dilate when it's dark, and we shiver in cold temperatures but sweat when it's hot. Selye, who was studying rats, discovered that they produced similar bodily responses across a wide range of stressful situations (Sapolsky 1998). Of course, the specific situations were different for laboratory rats than they would be for humans: being given repeated injections, being accidentally dropped on the floor, being chased with a broom, and so on. (Selye was a rather clumsy experimenter in his early days!) However, all of these events seemed to create the same set of physiological reactions in the rats.

Selye had identified a programmed set of responses that occur in animals when they're under stress. These responses are characteristic of many animals, including birds, reptiles, and mammals. Humans often like to think of themselves as superior to animals, but in terms of programmed responses we operate in much the same way as other vertebrate species; we have similar programmed physiological responses that allow us to react quickly in dangerous situations. Whether we're being chased by a bear, being asked to dance at a party, or being fired from a job, our bodies react in a way that's surprisingly similar to how the body of a rat being chased with a broom reacts.

Now, many decades later and thanks to extensive neurophysiological research, this reaction, which Selye dubbed the *stress response*, has been traced back to the central nucleus of the amygdala. The stress response produces a predictable set of physiological changes, including increased heart rate and blood pressure, rapid breathing, dilated pupils, a sudden increase of blood flow to the extremities, slowed digestion, and increased perspiration. All of these changes result from activation of the sympathetic nervous system and the release of stress hormones, such as cortisol and adrenaline. The fight, flight, or freeze response is a specific, acute, and intense form of the stress response. These physiological changes are hardwired into us, meaning we don't have to learn them. As discussed in part 1 of the book, they're very useful in escaping from danger, and many of our ancestors were probably saved by these swift and automatic responses that allowed them to escape the jaws of a predator or fight off an enemy.

Now, add to this response the fact that the amygdala is capable of identifying a situation as dangerous in just a fraction of a second, before the rest of the brain even knows exactly what the situation is. Other types of processing, such as perception, thinking, and retrieving memories from the cortex, may take more than a second to occur. You can see the significant advantages of being able to subconsciously identify whether a situation is dangerous or safe and respond accordingly before processing of the situation by other parts of the brain is completed. It could save your life! Consider Jason, who was crossing the street with his young daughter during winter when an oncoming car hit a patch of ice. It couldn't stop and slid dangerously close to them. Without thinking, Jason swiftly grabbed his daughter and leapt out of the path of the car—before he even realized he was doing so.

In order to operate quickly and automatically enough to be effective, the stress response *cannot* be based on the higher-level thinking processes we humans are so proud of possessing. It must operate more quickly than cortex-based circuitry would allow—if not, it may be too late!

Exercise: Recognizing the Stress Response in Your Anxiety Reactions

Which of the following experiences occur when you're feeling anxious? Read through the list and check any that apply to you:

_____ *Pounding heart*

_____ *Rapid breathing*

_____ *Stomach distress*

_____ *Diarrhea*

_____ *Muscle tension*

_____ *Desire to flee or withdraw*

_____ *Perspiration*

_____ *Difficulty focusing*

_____ *Immobilization*

_____ *Trembling*

All of the symptoms above can be traced back to activation of the stress response discovered by Selye. You may wonder why it's important that you recognize these symptoms as being connected to the fight, flight, or freeze response. A key reason is that they can be involved in a feedback loop in which they heighten anxiety. Many people who struggle with anxiety misinterpret these reactions as an indication that something negative is happening or going to happen. When they feel their heart pound, they might mistakenly believe that they're having a heart attack. Alternatively, they may be convinced that these sensations indicate that danger is imminent. But in reality, the symptoms they're experiencing are completely normal and simply mean the amygdala has been activated.

The stress response is essential in preparing us to respond immediately to emergency situations. Unfortunately, it isn't always useful in responding to the threats we face today. Increased heart rate, perspiration, and blood flow to your extremities aren't particularly useful when your boss tells you to increase productivity or face termination. They won't be helpful if you receive an overdue notice on your mortgage payment or your teenage daughter starts arguing with you. But these physiological reactions are hardwired into you, and once the central nucleus activates them, you'll have to contend with them.

The Role of the Central Nucleus of the Amygdala

The central nucleus of the amygdala is like an ignition switch. Once this tiny portion of the amygdala receives a signal from the lateral nucleus that indicates danger, it activates the stress response, and does so by sending messages to many other parts of the brain—making the amygdala a very well-connected player in brain processes. One of the most important parts of the brain it's connected to is the hypothalamus, a peanut-sized region of the brain that controls a variety of bodily processes, including metabolism, hunger, and sleep.

Because of its connection to the hypothalamus, the central nucleus can initiate the release of adrenaline, a hormone that increases heart rate and blood pressure, and cortisol, a hormone that causes glucose to be released into the bloodstream for quick energy. It can also activate the sympathetic nervous system, which can make swift changes in various physiological systems to allow us to react quickly, without conscious awareness or control. The brain is organized in such a way that the processing in the amygdala pathway occurs in milliseconds. Many studies conducted on rats and mice, which share this same stress response system, have greatly increased our understanding of these amygdala processes (LeDoux 1996).

One thing this research has clearly revealed is that when the stress response is activated, the amygdala's signals can influence and dominate brain functioning at all levels, something described by Joseph LeDoux (2002, 226) as a "hostile takeover of consciousness by emotion." We know it can be discouraging to learn that your clearest thinking skills and personal insights are essentially disabled by ancient brain structures that create fear-related responses. It's frustrating to realize that, at times, your insightful cortex can be entirely overtaken by your amygdala. But once you have this knowledge, you can use it. The key is to realize that many cortex-based coping strategies, such as telling yourself to stop being afraid or that there's no logical reason to be anxious, won't stop the activation of the stress response once it's initiated. At these times, strategies that target the amygdala are called for instead. The remaining chapters in part 2 will explain these approaches in detail.

When Panic Attacks

Undoubtedly the most unpleasant overactivation of the stress response is the panic attack. Panic attacks, a common difficulty faced by many people with anxiety disorders, also have their roots in activation of the central nucleus. These episodes of extreme agitation, or sometimes terror, fury, or immobilization, are accompanied by a pounding or racing heart, sweating, increased respiration, and often trembling or shaking. People experiencing a panic attack may feel the desire to attack someone (fight), an overwhelming urge to flee (flight), or an inability to take any action (freeze). Other possible symptoms include sympathetic nervous system reactions such as light-headedness, nausea, numbness or tingling, tightness in the chest, a sensation of being smothered, difficulty swallowing, or hot flushes or chills. In addition, the pupils dilate, making the world seem unnaturally bright, and time may seem to pass more slowly.

Few experiences in life are as unpleasant and overwhelming as a panic attack. Indeed, panic attacks are so distressing that some people fear that they're losing control, going crazy, or about to die. Symptoms usually last from one to thirty minutes but can return in waves and are not only frightening, but also quite exhausting.

Panic attacks typically occur when the amygdala responds to a cue or trigger you may not even be aware of. Basically, a panic attack is your body launching into the fight, flight, or freeze response at an inappropriate time due to an overreaction by the amygdala, often in response to some sort of trigger that poses no real danger. Of course, if there were some sort of real danger, you would need the physical responses you're experiencing, which would help you hide, run, or do battle, so these physical reactions wouldn't be overblown.

The central nucleus can cause a panic attack without any involvement of the thinking areas of the cortex, so from the cortex's perspective, panic attacks often seem to occur out of the blue. However, because the amygdala is reacting to some sort of trigger when it initiates a panic attack, it's common for people to have panic attacks repeatedly in the same or similar places, such as in a crowd, while driving, at church, or in a store. Although the trigger may be difficult to pinpoint, something has activated the amygdala and initiated a panic attack.

Most people have one or two panic attacks in their lifetime, and for most people these incidents are just a frightening inconvenience. People who experience repeated panic attacks are often diagnosed with panic disorder. When people begin to anticipate and fear having panic attacks and start avoiding places where panic attacks have hit them in the past, they've developed symptoms of *agoraphobia*—a fear of experiencing the fear that arises in situations in which they feel unable to escape. In this extremely debilitating condition, numerous places seem to be unsafe. By avoiding situations that might elicit panic, people with agoraphobia shrink their world in a misguided attempt to protect themselves. Agoraphobia has the potential to confine people to their home, or even to just one room, if it gets out of control.

The tendency to have panic attacks is at least partially due to genetics, and the search for the specific genes involved has begun (Maron, Hettema, and Shlik 2010). So some people have inherited a tendency for the amygdala to react in this way. In addition, panic attacks can also be caused by significant life changes or stresses, such as graduation, job changes, a death in the family, getting married or divorced, and other transitional events. Most people who experience panic attacks are women, but that statistic may be due in part to underreporting of panic attacks by men.

Some people who have panic attacks try to cope in unhealthy ways, such as by drinking alcohol or using drugs. These strategies may put a bandage on the problem, but they don't change the underlying circuitry of the brain in a helpful way. But don't despair! Even if you've inherited a reactive amygdala that's prone to panic attacks, you can use the language of the amygdala to gain control over your panic.

Exercise: Assessing Whether You've Experienced a Panic Attack

The following list can help you identify whether you've had a panic attack. If you've experienced many of the responses in this list at the same time, you were probably having a panic attack. At that time, you may not have recognized the experience for what it was: an extreme reaction resulting from the central nucleus activating the sympathetic nervous system and triggering the release of adrenaline. As you consider the following list of symptoms, you'll clearly see the influence of the sympathetic nervous system:

Pounding or racing heart

Feelings of panic or terror

Sweating

Hyperventilation

Light-headedness or dizziness

An urge to flee

Trembling or shaking

Nausea

Numbness or tingling

An urge to attack

Needing to go to the bathroom

Chills or hot flushes

Feelings of paralysis

Tightness or discomfort in your chest

A feeling of unreality

Difficulty swallowing

Fear of going crazy

Shortness of breath

Helping Your Amygdala Get Past Panic

You may wonder what the best way to cope with a panic attack is. If you're suddenly in the midst of a panic attack, there are three amygdala-based coping strategies that will work to calm you down: deep breathing, muscle relaxation, and exercise. They can't immediately shut down all of the activation that's been created in your body, but they will reduce your discomfort and shorten the duration of the panic attack.

Deep breathing: One of the best things to do when you're having a panic attack is to breathe slowly. Some of the symptoms of panic attacks, such as tingling or dizziness, are directly related to hyperventilation, or breathing too quickly. Taking slow, full, deep breaths that extend your chest and diaphragm outward is a good start. (The diaphragm is a muscle that extends across the torso beneath the lungs.) Slow breathing has been shown to decrease activation in the amygdala. We'll discuss breathing techniques in more detail in chapter 6.

Muscle relaxation: The amygdala is responsive to muscle tension, and tight muscles seem to increase amygdala activation. Learning and diligently practicing muscle relaxation techniques will help you both shorten panic attacks and make them less likely. We'll also discuss muscle relaxation techniques in more detail in chapter 6.

Exercise: We encourage you to pace or exercise during a panic attack. This will burn off the excess adrenaline that's in your system and should help shorten the panic attack. Remember, your body is prepared to fight or flee, so physical exertion is exactly what your body is ready to do. In chapter 9, we'll discuss the benefits of exercise in more detail.

One final and extremely important point: When you feel panicky, it's important to resist the strong urge to flee the situation. While it's an extremely frightening and unpleasant experience, a panic attack *won't physically hurt you*. In fact, the sensations you're experiencing are signs of a healthy, reactive body. Escaping the situation may make you feel better in the short term, but in the long run it will reinforce the power of panic attacks and make them more difficult to overcome. If possible, try to relax, breathe deeply, and remain in the situation. While that's definitely easier said than done, it's essential to gaining some control over your amygdala, since the amygdala learns from experience. If you leave the situation, your amygdala will learn to escape the situation rather than learning that the situation is safe. This point can't be overemphasized, and we'll return to it in chapter 8.

Helping Your Cortex Get Past Panic

Your cortex can't directly create a panic attack; it takes the amygdala and other brain structures to set the process in motion. But the cortex can definitely create the conditions for a panic attack or worsen a panic attack once it occurs. Sometimes the thoughts that people are having in their cortex put them more at risk for the amygdala to produce or worsen a panic attack. Therefore, the following cortex-based coping strategies may be helpful, especially before a panic attack really sets in.

Remember that it's only a feeling (albeit a very intense one): When the fight, flight, or freeze response is activated and you experience physical symptoms, the cortex's interpretation of those symptoms may cause anxiety to spiral out of control. If you think the symptoms mean you're having a heart attack, are going to lose control of yourself, or are going crazy, this will only worsen the panic attack. Recognizing that you're experiencing a panic attack and nothing more, and not buying into the cortex's misinterpretation of your amygdala-based symptoms, will help you recover more quickly.

Don't focus on panic attacks: One of the best ways to avoid having a panic attack is to stop worrying about panic attacks. Being preoccupied with panicking and constantly anticipating whether or when or where a panic attack may occur makes another panic attack more likely. Therefore, it's essential to keep your cortex from devoting too much thought to panicking or even the symptoms of panicking. When you're anxious, focusing on a bodily sensation, like sweating palms or a thumping heart, may lead to further anxiety-provoking thoughts that can build into a panic attack.

Distract yourself: Distraction is another cortex-based tool to use against a panic attack. Because the cortex can make a panic attack worse by focusing on symptoms, try to think about something— anything—other than panicking. (In chapter 11, we'll provide more guidance on using distraction.)

Let go of concerns about what others are thinking: People who have panic attacks often believe that everyone is looking at them or that they'll somehow embarrass themselves. If you feel symptoms of panic, try not to let your cortex guess what other people may be thinking. Other people probably won't be aware of what you're experiencing or won't care. Worrying about what other people think only creates additional stress when you're already having one of the most uncomfortable stress reactions possible.

Although the preceding tips may be helpful for preventing panic attacks, the efficacy of cortex-based approaches is limited once a true panic attack begins. In a full-blown panic attack, you're likely to be too anxious to think clearly because the amygdala is taking charge and shutting out the influence of your cortex. At such times, the only solution is to breathe slowly, try to relax, and distract yourself as you wait for the attack to pass. The good news is, it always passes. If others are present, the best way for them to help is to remind you to breathe deeply and relax your muscles, which will be naturally tensing and tightening as your adrenaline surges. If someone can help you use relaxation strategies, you'll probably be surprised at how much more quickly your level of panic decreases.

Under no circumstances should you listen to people who tell you that it's all in your head or you should just get over it. Panic attacks are caused by an overreaction of the amygdala. They're a biological reality, and you can't reason yourself out of them using your cortex.

Once the central nucleus initiates a panic attack, you need to use the coping strategies mentioned in this chapter, which we'll discuss in greater depth in chapters 6 and 9. They will help you get through the panic attack. It may be an extremely uncomfortable experience, but remember, you aren't in danger and the panic isn't actually causing you any harm.

Don't Freeze: Retraining the Amygdala to Resist Avoidance

If your amygdala seems to be wired to produce freezing or immobilization rather than the more active fight or flight responses, you're particularly at risk to become withdrawn and avoidant. This can spiral into crippling tendencies or even agoraphobia—that fear of fear mentioned before, which can severely restrict your life. To minimize this tendency, it's necessary to engage in active rather than passive responding.

Research (LeDoux and Gorman 2001) has shown that it's possible to avoid activating the pathways responsible for the freeze response, which lead from the central nucleus of the amygdala to the brain stem at the back of the brain, atop the spine. Doing so requires diverting the flow of information leaving the lateral nucleus of the amygdala. Rather than traveling to the central nucleus, information from the lateral nucleus is instead sent to the basal nucleus of the amygdala, which promotes active responses.

Switching to this alternative pathway requires using an *active coping strategy* (LeDoux and Gorman 2001). When you feel caught in the grip of the freeze aspect of the fight, flight, or freeze response, engaging in active coping strategies can rewire your amygdala to stop choosing a passive response. Initially, it's important to just do *something*—anything at all. You may not feel capable of doing complicated or demanding tasks, but don't allow yourself to freeze and become immobilized, like a frightened rabbit, because this will strengthen the wiring that underlies passive responding. Find something active you can do, even if it's just calling someone on the phone. In fact, social activities that involve enjoyable interactions with others, or simple pleasurable activities that distract you from your worries, can keep your amygdala from creating responses characterized by freezing, avoidance, and immobilization.

Consider Patricia, who often felt too panicky to go to work and therefore remained at home immobilized many mornings. She usually stayed in bed, and she felt it was somehow wrong for her to do anything enjoyable, since she wasn't able to go to work. But when she started being more active at these times, calling friends or family members or doing something simple but enjoyable like working on a jigsaw puzzle, she found that she was often able to go to work after all, albeit late. She was shifting her amygdala to a more active response, making her less likely to engage in avoidant behavior for the rest of the day.

Summary

You've learned about the nature and purpose of the stress response and its most intense version, the fight, flight, or freeze response. Hopefully you understand the importance of not interpreting a stress response to mean that actual dangers, either physical or external, are present. While this response is intrinsically distressing, especially when it comes in its most extreme form—a panic attack—you now have new ways of thinking about it and counteracting it. You also know that active responding is necessary to overcome the inclination toward avoidance. Whether your amygdala tends to use aggressive responding (fighting), avoidant responding (fleeing), or passive responding (freezing), you can teach it alternatives. Knowledge that the amygdala can be trained to respond in more beneficial ways is very powerful. In the chapters that follow, you'll learn how to use a variety of strategies, including relaxation (chapter 6), exposure (chapter 8), and exercise (chapter 9), to help your amygdala respond in new ways, giving you more control over your life.

CHAPTER 6

Reaping the Benefits
of Relaxation

Both in your daily life, as well as when you are actively engaged in the exposure techniques described in chapters 7 and 8, we believe you will find relaxation practices extremely valuable in reducing anxiety. When you feel anxious, other people may try to help you feel better by telling you not to worry, that everything will be all right, or that you have no reason to be anxious. You may try the same strategy with yourself. The problem with this approach is that when you try to use thinking processes and logic to cope with feelings of anxiety, you're relying on cortex-based methods. And by itself, the cortex can't reduce the stress response, for two primary reasons. First, as we've noted, the cortex doesn't have many direct connections to the amygdala. Second, the initiator of the stress response is the amygdala. Therefore, interventions that target the amygdala are more direct and effective in easing anxiety.

By activating the sympathetic nervous system (SNS) and stimulating the release of adrenaline and cortisol, the central nucleus can instantly increase heart rate and blood pressure, direct blood flow to the extremities, and slow digestive processes. Consider Jane, who had to give a speech. She found herself trembling, with her heart pounding and her stomach feeling queasy. These spontaneously activated processes, whether described as anxiety, the stress response, or the fight, flight, or freeze response, result from brain activities that don't lie within conscious awareness.

However, lack of conscious awareness doesn't mean we completely lack control over these processes. For example, although we don't consciously control our rate of breathing most of the time, we can deliberately modify it if we choose to do so. A variety of techniques have been developed for activating the parasympathetic nervous system (PNS), which reverses many of the effects the central nucleus creates by activating the SNS. As mentioned in chapter 1, whereas activation of the SNS creates the fight, flight, or freeze response, the action of the PNS is often referred to as "rest and digest." It slows heart rate and increases secretion of gastric juices and insulin as well as activity of the intestines.

The PNS is more likely to be activated when people are relaxed. That's why medical professionals often encourage anxious patients to engage in activities that strengthen the tendency toward PNS activation and decrease SNS activation. Relaxation training is one of the primary methods suggested to facilitate PNS activation. A variety of studies have shown that techniques that promote relaxation, such as breathing exercises and meditation, reduce activation in the amygdala (Jerath et al. 2012). When you reduce amygdala activation, you reduce SNS responding, and with practice, the PNS can be trained to intervene.

Relaxation Training

Relaxation training has been formally recognized since the 1930s, when physician and psychiatrist Edmund Jacobson (1938) developed a process called progressive muscle relaxation. Recent neuroimaging studies have identified actual changes in the brain that occur when people practice various relaxation techniques, including meditation (Desbordes et al. 2012), chanting (Kalyani et al. 2011), yoga (Froeliger et al. 2012), and breathing exercises (Goldin and Gross 2010). These studies have found that many of these approaches almost immediately reduce activation in the amygdala, which is good news for people who struggle with anxiety. We present several such techniques in this chapter, and we encourage you to try all of them to discover which ones work best for you or which you prefer. Whichever you choose to practice in the long run, you'll know that scientific evidence indicates that you can directly affect your amygdala when you use them.

Most approaches to relaxation focus on two physical processes: breathing and muscle relaxation. Individuals respond in different ways to various relaxation strategies, but virtually everyone will benefit from relaxation training. Relaxation is a very flexible approach that can be used in many situations, and it has many beneficial effects, especially in the short term. The effectiveness of relaxation strategies is often immediately apparent. Relaxation is also an integral component of more complex approaches to reducing stress and anxiety, such as meditation and yoga.

Breathing-Focused Strategies

If you take a few moments right now to attend to your breathing, you may be able to demonstrate to yourself some of the basic effects of relaxation. Take a deep breath, making a point of expanding your lungs as you inhale deeply and slowly. Don't hold your breath. Allow yourself to exhale naturally. Some people feel a reduction in anxiety almost immediately when they do this for several minutes. Merely altering your breathing and adopting a slow rhythm of deep breathing can be soothing and relieve stress.

People tend to hold their breath or breathe shallowly when experiencing something stressful, without being aware that they're doing so. Several specific breathing techniques can help you consciously deepen your breathing and reduce your heart rate to counter physiological processes that are part of SNS activation. Here are a few that are especially effective.

Exercise: Slow, Deep Breathing

The first technique is basically the same as what we described above: slow, deep breathing. Practice it now, taking a few deep breaths. Inhale slowly and deeply, and exhale fully. Don't force your breathing; rather, breathe gently both in and out. It doesn't matter whether you breathe through your mouth or nose—just breathe in a comfortable manner. Note how this deliberate slowing and deepening of your breathing affects you. Does it have a calming effect?

Not everyone finds slow, deep breathing to be calming. Increased attention to breathing can increase anxiety in some people, especially those with asthma or other breathing difficulties. In such cases, people may get greater benefit from relaxation strategies that focus on reducing muscle tension or that use music or movement. That said, most people are surprised at how effective simple breathing exercises can be in reducing anxiety and increasing calmness almost immediately. Many students find this approach helpful before and during exams. Nervous drivers use it while on the road, and people who are claustrophobic often find it helpful when they're in an enclosed space. Plus, the breath is readily available in all situations. You can practice slow, deep breathing almost anytime and anywhere, and it's completely free!

Breathing Techniques to Counter Hyperventilation

When people are anxious, they're likely to breathe quickly and shallowly. They may not get enough oxygen, which produces an uncomfortable sensation. Hyperventilation can also result, due to expelling carbon dioxide too quickly, resulting in low levels of carbon dioxide in the blood. This can cause dizziness, belching, a feeling of unreality or confusion, or feelings of tingling in the hands, feet, or face.

Hyperventilation disrupts the balance between oxygen and carbon dioxide in the body, and the amygdala detects this instantly. Correcting this imbalance using deliberate breathing techniques sends a signal to the amygdala to relax. Consider Toni, who thought her feelings of dizziness and tingling were just part of her anxiety.

When she learned that she was experiencing the results of hyperventilation, she found that she could reduce those symptoms by simply attending to her breath.

People who are hyperventilating are often instructed to deliberately slow their breathing or breathe into a paper bag. The bag captures carbon dioxide when they exhale; therefore, breathing in from the bag increases the amount of carbon dioxide inhaled and replaced in the bloodstream. It's a very effective method of reversing light-headedness and other anxiety symptoms.

Diaphragmatic Breathing

A specific method of breathing known as *diaphragmatic* or *abdominal breathing* is recommended for its particular effectiveness in activating the PNS (Bourne, Brownstein, and Garano 2004). This type of breathing helps turn on a relaxation response in the body. In this technique, you breathe more from the abdomen than from the chest, and the movement of the diaphragm (the muscle under the lungs) has a massaging effect on the liver, the stomach, and even the heart. This type of breathing is thought to have beneficial effects on many internal organs.

Exercise: Diaphragmatic Breathing

To practice diaphragmatic breathing, sit comfortably and place one hand on your chest and the other on your stomach. Take a deep breath and see which part of your body expands. Effective diaphragmatic breathing will cause your stomach to expand as you inhale and retract as you exhale. Your chest shouldn't move much at all. Try to focus on breathing deeply in a manner that expands your stomach as you fill your lungs with air. Many people tend to pull their stomachs in as they inhale, which keeps the diaphragm from expanding downward effectively.

Shifting Breathing Patterns with Regular Practice

Healthy breathing techniques can become second nature with practice. Pay attention to your style and pattern of breathing and work to consciously modify it. Practicing for brief, five-minute sessions at least three times a day can increase your awareness of your breathing habits and help you train yourself to breathe in more healthful, effective ways.

Also try to notice times when you're holding your breath, breathing shallowly, or hyperventilating, and then make a deliberate effort to adopt a better breathing pattern. Breathing is an essential bodily response that you can control, and in the process, you can reduce amygdala activation and its effects. With practice, you'll find that healthy breathing becomes a valuable tool and that it alleviates many symptoms that you may have thought were part of your anxiety.

Muscle-Focused Relaxation Strategies

The second component of most relaxation training programs is muscle relaxation, which also works to counter amygdala-based activation of the SNS. The SNS creates increased muscle tension because fibers in the SNS activate muscles in preparation for responding. Although the problems we face in today's world are seldom things we can fight or run from, this muscle tension is programmed into the nervous system, and people often feel stiff and sore because of it. Luckily, as with breathing, you can modify your muscle tension if you deliberately attend to it. Additionally, relaxing your muscles can promote the PNS responding you want to increase.

People are often completely unaware that muscle tension builds up as a result of amygdala-based anxiety. However, if you observe yourself, you may find that you often clench your teeth or tense your stomach muscles for no apparent reason. Certain areas of the body seem to be vulnerable as repositories for muscle tension, including the jaw, forehead, shoulders, back, and neck. Constant muscle tension uses energy and can leave people feeling tight and exhausted at the end of the day. The first step in reducing muscle tension is to discover which areas of your body tend to tighten up when you're anxious. The next exercise will help you do just that.

Exercise: Doing a Muscle Tension Inventory

Right now, check your jaw, tongue, and lips to see if they're relaxed or tense. Consider whether muscle tension is tightening your forehead. Determine whether your shoulders are loose, low, and relaxed, or tightened up toward your ears. Some people tense their stomach as though they expect to be punched any moment. Others clench their fists or curl their toes. Take a brief inventory of your entire body to see where you're holding your tension at this moment.

Once you have an idea of which areas in your body are vulnerable to muscle tension, you're ready to learn to relax those areas. To begin, you may find it helpful to experience the difference between feelings of tension and relaxation in your muscles. The next exercise will help you explore that.

Exercise: Exploring Tension vs. Relaxation

Tension is often experienced as a tight or strained feeling. In contrast, relaxation is often described as a loose and heavy feeling. To help you tune in to your own experience of tension versus relaxation, make a fist with one of your hands and clench it tightly while counting to ten. Then let that hand relax by dropping it limply into your lap or onto another surface. Compare the feeling of tension that you experienced as you clenched your fist to the feeling of relaxation while the muscles are loose and limp. Do you recognize a difference? Also compare the hand that you tightened and relaxed to the other hand and notice whether one hand feels more relaxed than the other. Often, tensing and releasing muscles helps create a feeling of relaxation in those muscles.

Exercise: Progressive Muscle Relaxation

One of the most popular muscle relaxation techniques is *progressive muscle relaxation* (Jacobson 1938), which involves focusing on one muscle group at a time. It's a practice of briefly tensing and then relaxing the muscles in one group, then switching to the next muscle group, and then the next until all major muscle groups are relaxed. When you first learn progressive muscle relaxation, it may take you up to thirty minutes to complete the entire process of tensing and relaxing every muscle group. With time and practice, you can train yourself to relax your muscles more readily so much less time is required. If you practice diligently, eventually you'll probably be able to achieve a satisfying level of relaxation in less than five minutes.

We recommend doing this exercise while sitting in a firm chair. Begin by focusing your attention on your breathing. Take a few moments to practice slow, deep, diaphragmatic breathing. If you can slow your breathing to five or six breaths per minute, it will promote relaxation. You may find it helpful to think a word, "relax" or "peace," as you breathe. Or you might prefer to use imagery to enhance relaxation, perhaps imagining that with each exhalation you're breathing out stress and with each inhalation you're breathing in clean air. Consider imagining that the stress has a color (perhaps black or red) and that you're breathing it out and filling yourself with stress-free, colorless air.

Next, you'll begin to focus on specific muscle groups. Throughout, maintain some attention on your breath and keep it slow and deep.

Begin by tensing the muscles in your hands by briefly clenching your hands into fists. After a few seconds, let go and try to completely relax your hands, including each finger. Let your hands drop into your lap and feel gravity pulling them down. You may need to wiggle your fingers to relax them.

Next, focus your attention on your forearms and create tension by making fists again and also tightening your forearm muscles to briefly create muscle tension in your forearms. After just a few seconds, drop your hands into your lap and allow the muscles in your hands and forearms to completely relax. Focus on releasing any tension in your forearms and feeling the heaviness of relaxation.

Next, move to your upper arms, pulling your hands and forearms close to your upper arms and tensing your biceps. Then completely loosen and relax, allowing your arms to hang at your sides and feeling how the weight of your relaxed hands and arms lengthens your biceps into a relaxed state. Shaking your arms may help release any remaining tension.

Now turn your attention to your feet and tense them by curling your toes. After a few seconds, release the tension by wiggling or stretching your toes. Continue working up through your legs in the same way. Tense your calves by leaving your heels on the ground and flexing your feet and toes upward, then relax by stretching your feet out comfortably. Tense your thighs by pushing your feet into the ground, then release and focus on the sensations of relaxation. Then tense and release your buttocks.

Now move to the muscles in your forehead and tense them by frowning. To relax, lift your eyebrows, then allow them to relax into a comfortable position. Next, turn to your jaw, tongue, and lips, clenching your teeth together firmly, pushing your tongue against your teeth, and pushing your lips together. Release the tension in your mouth by allowing it to be slightly open, with your lips and tongue relaxed. This is a good time to check to make sure your breathing is still slow and deep.

Now tense your neck by tipping your head back. To relax, gently tip your head to one side, then the other, then gently tip your chin toward your chest. Next, tense your shoulders by bringing them up toward your ears, then relax completely, allowing the weight of your arms and hands to pull your shoulders down. Finally, turn to your torso and tighten the muscles in your abdomen as though bracing for a punch to the stomach. Then relax completely, allowing your stomach muscles to be loose and soft.

Take a moment to feel the sense of deep relaxation throughout your entire body, then gently stretch comfortably and return to other activities.

* * *

We recommend that you practice progressive relaxation daily, preferably at least two times per day, until you've reduced the time it takes to achieve relaxation to approximately ten minutes. Typically, people eventually learn to relax most of their muscles without having to tense them first, perhaps tensing only stubborn muscle groups that seem particularly vulnerable to stress-related tension. Different groups of muscles may be problematic for different people. For example, one person may find that he's constantly gritting his teeth, while another holds tension in her shoulders. Learning to relax effectively is an individual process that you must tailor to yourself, with your specific needs in mind.

Designing Your Own Strategies for Muscle Relaxation

Try a variety of approaches to muscle relaxation and choose the one that's most effective for you. After all, you know yourself best. As you experiment with different approaches, do bear in mind that, with any technique, more practice is often required at first.

If you have an injury or chronic pain difficulties, tensing your muscles may be counterproductive. If this is the case for you, you can follow the above procedure for progressive muscle relaxation, but instead of tensing each muscle group first, simply turn your attention to each muscle group in turn and try to completely relax and loosen all of the muscles in that group. Even if you use the tensing recommended in progressive muscle relaxation, once you master the process of relaxing your muscles you should feel free to use the tension-free approach, which is more efficient because it's quicker. For the most effective approach to reducing activation of the amygdala and SNS in order to produce a PNS response, combine breathing-focused methods with muscle relaxation.

Imagery

Using imagery, or visualization, is also a beneficial relaxation strategy. Some people have the ability to imagine themselves in another location and can use visualization to effectively attain a relaxed

Reaping the Benefits of Relaxation

state. If you're one of those individuals, you may find that imagining yourself on a beach or in a peaceful forest glade allows you to achieve a more satisfying state of relaxation than a focus on muscle relaxation does. Either way, the most important goal is to achieve *deep breathing* and *relaxed muscles*. That's the key to reducing activation of the amygdala. The truth is, it doesn't matter whether you attain this state by directly focusing on your breathing and muscles or by imagining yourself in a setting that allows you to relax.

Exercise: Assessing Your Ability to Use Imagery

Read through the following description of a relaxing situation, then take a few moments to close your eyes and imagine yourself in that setting.

> Imagine yourself on a warm beach. Feel the sun warming your skin and the cool breeze coming off the water. Listen to the sounds of the waves as they wash against the shore and the calls of birds in the distance. Allow yourself to relax and enjoy the beach for several minutes.

How well were you able to imagine yourself in the described setting? If the visualization arose for you readily and you find it pleasant and engaging, we highly recommend that you use imagery as one of your relaxation strategies. It may allow you to achieve a relaxed state more effectively than other approaches. On the other hand, if you found it difficult to relax using this method and noticed your mind wandering, you'll probably find other strategies more helpful.

Exercise: Practicing Imagery-Based Relaxation

When you use imagery to relax, you take yourself to another location in your imagination. Start by slowing your breathing and relaxing your body as you mentally travel to another scene. We've provided a guided script based on the image of a beach below, to give you an overview of the process, but feel free to choose any location you enjoy. The key is to close your eyes and allow yourself to experience this special place in detail. Try to use all of your senses (sight, sound, smell, touch, and

105

even taste) as you imagine yourself in this particularly relaxing situation. You might ask someone to read this script to you so you can close your eyes and focus.

> *Imagine yourself walking on a sandy path to a beach. As you walk on the path, you're surrounded by trees that keep you in dark shade. You feel the sand begin to get into your shoes as you walk along. You can hear the leaves in the trees softly moving in the wind, but up ahead you hear another sound: gentle waves washing up on shore.*
>
> *As you continue, you leave the shade of the trees to walk out onto a sunny, sandy beach. The sun warms your head and shoulders as you stand still for a moment to take in your surroundings. The sky is a beautiful shade of blue, and wispy white clouds seem to hang motionless in the sky. You take off your shoes and feel the warm sand as your feet sink in. Holding your shoes, you move toward the water. The sound of the waves rhythmically washing up on the shore has a hypnotic quality. You breathe deeply, in unison with the waves.*
>
> *The water is dark blue, and far off, on the horizon, you can see a darker blue line where the water meets the light blue sky. In the distance, you see two sailboats, one with a white sail and one with a red sail; they appear to be racing one other. The damp smell of driftwood reaches your nose, and you see some driftwood nearby. You place your shoes on a smooth, weathered log and walk toward the waves.*
>
> *Seagulls swoop overhead, and you hear their excited cries as they glide on the gentle breeze coming in with the waves. You feel the breeze on your skin and smell its freshness. As you walk toward the waves, you see the sun reflected on the water. You walk into the damp sand, leaving footprints now as you walk along the shore. A wave breaks over your feet, surprisingly cold at first.*
>
> *You stand still as the waves wash over your ankles. Listening to the repetitive sound of the waves and the cries of the gulls, you feel the wind blowing your hair away from your face. You take slow, deep breaths of the cool, clean air...*

We recommend that you end each imagery session gradually, counting backward slowly from ten to one. With each number, gradually become more aware of your surroundings—the actual environment around you. When you reach one, open your eyes and return to the present moment feeling refreshed and relaxed.

Through imagery, you can take a trip each day that's limited only by your imagination and that can decrease SNS activation in just a few minutes. Choose locations that you can explore and that lead to feelings of peace and comfort. As you practice, remember that visualization will be most effective at reducing amygdala activation if you achieve relaxation in your muscles and slow and deepen your breathing.

Meditation

Various meditative practices—including mindfulness, which is currently the most popular approach—have been shown to reduce amygdala activation (Goldin and Gross 2010). All forms of meditation involve focusing attention, perhaps on the breath, or perhaps on a specific object or thought. Extensive research on meditative practices has shown that they affect a variety of processes in both the cortex and the amygdala (Davidson and Begley 2012). Because it's a relaxation strategy that can target the cortex, we'll provide a more detailed explanation of meditation, and mindfulness in particular, in chapter 11, "How to Calm Your Cortex." However, meditation is also an effective method for calming amygdala activation, particularly when the focus of attention is the breath.

If you're experienced in meditation or interested in it, we encourage you to pursue this practice. Research has demonstrated that a regular practice of meditation can reduce a variety of stress-related difficulties, including high blood pressure, anxiety, panic, and insomnia (Walsh and Shapiro 2006). But most importantly for people who struggle with anxiety, meditation has also been shown to have direct and immediate calming effects on the amygdala. It produces both short-term and long-term effects in the amygdala, reducing amygdala activation in a variety of situations and increasing PNS activation (Jerath et al. 2012). Clearly, it's an effective relaxation strategy, and we've spoken to many people who find that incorporating regular meditation into their morning routine decreases their overall anxiety and helps them feel more prepared to cope with the demands of the day.

Breath-Focused Meditation

Many approaches to meditation include a focus on the breath, with meditators concentrating on the experience of breathing or modifying the breath in some way. Studies have shown these breath-focused practices to be effective in reducing the amygdala's reactivity. In one study (Goldin and Gross 2010), people with social anxiety were trained in either breath-focused meditation or distraction techniques. Then they were presented with negative self-beliefs related to their anxiety, such as "People always judge me." Those who had engaged in breath-focused meditation had less amygdala activation in response to the statements. In another study (Desbordes et al. 2012), adults without an anxiety disorder were trained in breath-focused or compassion-focused meditation. All experienced a general and lasting decrease in amygdala activation, with those who were trained in breath-focused meditation experiencing greater benefits.

Using meditation effectively requires some practice. In most studies, people received at least sixteen hours of training prior to being assessed as to whether practicing meditation had changed their amygdala functioning. So for maximum benefit, you may wish to seek specific training from a therapist or other instructor. Mindfulness approaches to meditation are particularly popular at this time, and books on mindfulness techniques are plentiful. (We list some that we recommend in the "Resources" section.) There's also a good chance that you can find a therapist or other mindfulness meditation instructor in your area.

Meditation techniques that focus on breathing and relaxation seem to be most effective in modifying the amygdala's response. One study (Jerath et al. 2012) found that after meditation, people have a slower breathing rate and increased PNS activation. These effects are probably central to its effectiveness. The next exercise will help you enjoy the benefits of reducing activation of the amygdala through a focus on breathing.

Exercise: Breathing Meditation

This practice is very straightforward. Close your eyes if you like and simply focus your attention on your breath. Breathe in through your nose, and as you do so, attend to the way the air feels as it travels through your nostrils. Don't force the breath; simply take in long, slow breaths and observe the sensations of inhaling and exhaling in your nose and chest. Enjoy the sensations of breathing.

Notice the difference between the air moving into your nostrils and the air coming out. Pay attention to the way the air causes your lungs to expand. Notice the different stages of breath: as you inhale and air fills your lungs, and as you exhale and your lungs empty. Then focus only on the process of inhalation, noting that the beginning of an inhalation feels different than the process of inhalation or the end of an inhalation. Notice the same aspects of exhaling: the beginning, the middle, and the end.

During this meditation, your mind is likely to wander to other thoughts. This is common and natural. When this happens, just bring your focus back to your breath. If it wanders fifty times, bring it back to your breath fifty times.

Continue focusing on your breath for about five minutes, then slowly and gently come out of the meditation.

Relaxation as a Daily Process

Whatever approach you choose, working opportunities for relaxation into your daily schedule is an essential part of coping with fear and anxiety. Consider practicing in the morning or evening, during work breaks, or even on public transportation or while walking. Try to schedule at least three or four opportunities for some type of relaxation each day. Even a five-minute relaxation session can reduce your heart rate and muscle tension. If you're prone to panic attacks, relaxation strategies can help prevent them or provide relief. In addition, regular practice can help reduce your overall stress level.

Like most people who struggle with anxiety, you may find that tension tends to build gradually over the course of your day. You can thank your central nucleus and SNS for keeping your body in this tense, alert state. As your central nucleus activates your SNS during the day, you can keep switching your SNS off by using relaxation to activate your PNS. Just like an air conditioner that keeps cooling a home, you need to keep cooling off your amygdala. The advantage of the techniques in this chapter is that, unlike air-conditioning—or medication or psychotherapy—they cost nothing beyond a small amount of time. If you practice relaxation techniques routinely, eventually they will become second nature and help decrease your general anxiety level.

We've outlined a number of different approaches to relaxation that can be helpful in reducing activation of the amygdala. There's no single right way to achieve the relaxation that reduces amygdala-based anxiety; you simply need to find which techniques work best for you. Of course, the ability to relax is only beneficial if you use it when you need it, so be sure to choose strategies you can incorporate into your daily life. If you're only able to achieve muscle relaxation while lying down, or can only use imagery when your surroundings are perfectly silent, you won't be able to use those techniques in all situations. This may mean that you sometimes need to use different techniques, or it may just mean you need more practice.

Summary

Sometimes you may try to reason yourself into calming down, using cortex-based strategies in an attempt to think yourself into relaxing. We hope this chapter has helped you see the usefulness of another approach. Instead of focusing on your thoughts (the cortex approach), you can work directly on the physiological responses that the central nucleus of the amygdala is initiating and counter them with PNS activation. The ultimate goal is to increase activation of your PNS to help you recover from the stress response and promote well-being. Slower breathing and relaxed muscles will send a message directly to the amygdala that the body is calming down, which is more likely to calm the amygdala than all of the thinking you can do.

CHAPTER 7

Understanding Triggers

In this chapter, we turn our attention from the central nucleus of the amygdala, which *initiates* the stress response, to the lateral nucleus of the amygdala, which receives information from the senses and forms emotional memories. The lateral nucleus is the decision-making portion of the amygdala, which determines whether the central nucleus should react to a particular sight or sound. It does this by scanning the sensory information it's receiving and, based on emotional memories, determining whether a threat exists. The lateral nucleus also creates anxiety-related memories, and changing those memories is essential in rewiring the amygdala. In order to communicate with the lateral nucleus and influence the memories it creates, you need to have a clear understanding of the language of the amygdala.

Using the Language of the Amygdala

In chapter 2, you learned that the language of the amygdala is based on associations. Specifically, the lateral nucleus recognizes associations between events occurring in close proximity in time. We learn to fear triggers that are associated with negative events, whether or not a trigger actually causes a negative experience. When a trigger is paired with a negative event, the amygdala is programmed to produce anxiety. Consider Lynn, who experienced a sexual assault. She developed a strong panic reaction to the smell of the cologne her assailant had worn, even though the cologne wasn't itself relevant to the attack.

In the language of the amygdala, the pairing of a trigger with a negative event is very powerful. Thinking processes from the cortex, like logic and reasoning, are of little use when dealing with fear and anxiety in the amygdala. Trying to reason yourself out of anxiety isn't very effective because you aren't speaking the language of the amygdala. You need to learn to focus on pairings, and this chapter will help you learn how to do that.

Working with amygdala-based emotional memories can be difficult, as these memories may be formed and recalled by the amygdala outside of your awareness. Therefore, many of their influences occur without your conscious knowledge. A variety of sensory experiences—even seemingly irrelevant cues that you may barely notice in a situation, like a sound or a smell—can create anxiety. For this reason, it may take some work to learn to recognize triggers, because you may not be consciously aware of them.

Understanding Triggers

A trigger is an anxiety-provoking stimulus, such as a sensation, object, or event, that was originally neutral, meaning it wouldn't cause fear or anxiety for most people. Originally, it wasn't associated with any emotional memories, positive or negative, and therefore didn't cause any reaction.

In chapter 2, we discussed Don, a Vietnam War veteran whose PTSD is set off by the smell of a particular soap. For Don, the soap was associated with negative events, so he had a negative reaction to it. For Don's wife, however, the soap was neutral because her amygdala hadn't created any emotional memories about it. Therefore, the soap doesn't trigger any reaction in her.

Most sensations, objects, and events aren't typically associated with positive or negative emotions in most people. A crowd is just a crowd, an elevator is just an elevator, and so on. These things become triggers when an emotional memory is made, whether with anxiety, happiness, or even affection.

The reason a stimulus that's been paired with a negative event is called a trigger is because, due to association, it will cause, or trigger, a fearful reaction. This change is due to memories the lateral nucleus makes when the trigger is paired with the negative event. For example, in Lynn's case, the smell of a specific cologne didn't originally cause her to experience any fear or anxiety. It was just a neutral smell. But when Lynn was assaulted, her amygdala created an emotional memory about the cologne her assailant was wearing. This process is illustrated in figure 6.

Association

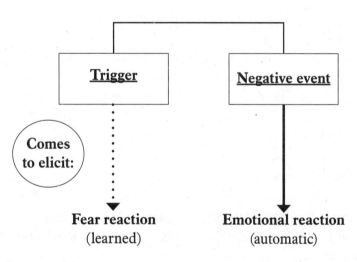

Figure 6. How triggers come to produce anxiety responses.

A formerly neutral trigger has been paired with an emotion-arousing negative event, meaning an event that results in discomfort, distress, or pain. As you can see in figure 6, a negative event leads to an emotional reaction. Lynn's experience of being sexually assaulted is obviously a negative event.

In the diagram, the line connecting the two boxes signifies a pairing or association between the trigger and the negative event. This is a visual reminder that the negative event occurs shortly after the trigger. The two are paired together, with the negative event following the trigger in time. Lynn smelled the cologne just before the sexual assault occurred, and that created a pairing. These kinds of pairings are very important to the amygdala.

The pairing between the trigger and the negative event changes the reaction caused by the trigger. Instead of eliciting no emotional reaction, the trigger now leads to a learned fear reaction. So in Lynn's case, because the cologne was paired with the sexual assault, the cologne now causes Lynn's amygdala to produce a fear reaction. Before, the cologne was neutral. Now, it's a trigger that will cause fear. This fear reaction was learned in the lateral amygdala and stored as an emotional memory.

Diagramming to Identify Triggers

Diagrams like figure 6 can be used to identify triggers. We'll use another example to demonstrate how this works. Normally, the sound of a car horn doesn't cause a strong panic reaction. For a person to learn to respond in this way to a horn, it must be paired with something very negative, like an accident. See if you can diagram the pairing that would occur in such a situation. (If you like, visit http://www.newharbinger.com/31137, where you'll find a downloadable file with the correct diagram, as well as a PowerPoint presentation that provides helpful guidance on diagramming the language of the amygdala. See the back of the book for more information about access.)

In this example, the pairing of the sound of a horn with an automobile accident causes the lateral amygdala to form a memory about car horns. Afterward, whenever the amygdala hears a car horn, it produces a fear reaction. It's important to bear in mind that the horn didn't cause the accident; it's just *associated* with the accident. Remember, the language of the amygdala is based on associations or pairings, not cause and effect.

Triggers can come in many forms. They may be sights, smells, sounds, or situations. For example, after a person is in an auto accident, the sight of a specific intersection, the smell of burned rubber, the sound of brakes, or even the sensation of braking can all cause the person to feel fear. In fact, after a single traumatic experience, a number of different triggers (intersection, burned rubber smell, the sound of brakes or a car horn, and the sensation of braking) can each cause fear. Each trigger becomes a cue for fear or anxiety.

The diagram in figure 6 is designed to help you remember the amygdala's learning process. You can remember the difference between the trigger and the negative event by noting the symbols that connect each stimulus to its response in the diagram. The bold arrow from the negative event to the emotional reaction indicates that there's an *automatic* connection between the negative event (such as an accident) and the reaction. In contrast, the connection between the trigger (such as a car horn) and the fear reaction is one that's created, or learned, by the lateral amygdala as a result of the pairing of the trigger and the negative event. The dotted line signifies that the fear response is a learned response, and what's learned can be changed.

Using Diagramming to Understand the Language of the Amygdala

Learning to identify triggers and the negative events they're associated with is very helpful in understanding the language of the amygdala and its role in producing your anxiety. Here are some helpful guidelines: Both the trigger and the negative event are stimuli, meaning they're objects, events, or situations that you see, hear, feel, smell, or experience. The trigger differs from the negative event because you *learn* to fear or be anxious about the trigger, whereas the negative event is something you don't have to learn to react to. The trigger activates emotions in you, even though you may know those emotions aren't logical, and even though you may want to stop responding to the trigger in that way.

Learned fear responses can occur with a variety of objects, sounds, or situations, as long as they're associated in time with a strong negative event. Feeling motion sickness on a roller coaster can cause someone to be fearful of amusement park rides. On the other hand, a different person might feel excited on the same ride and love the roller coaster as a result. The lateral nucleus of the amygdala recognizes and remembers these associations, and that's what determines our subsequent reactions. These emotional memories can be very strong and enduring.

Exercise: Diagramming Triggers

Take some time to learn to how to diagram triggers, negative events, and learned versus automatic responses, because this is the language of the amygdala. Knowing this language gives you the ability to communicate with the amygdala. In most cases, the trigger and negative event will be the only parts of the diagram that you need to figure out. A blank version of the diagram is available for download at http://www.newharbinger.com/31137, where you'll also find a few examples you can use to practice diagramming associations. (See the back of the book for information on how to access this content.) This will help you learn to identify triggers and distinguish them from negative events.

Your Amygdala Knows

The most powerful tool in coping with anxiety reactions is having an in-depth understanding of your own unique anxiety responses. To be effective in retraining your brain to resist anxiety reactions, specific knowledge about your own triggers is essential. For this reason, it's crucial that you take a close look at situations and events that are connected with your anxiety responses. This will help you identify the triggers that you need to address through exposure therapy, a powerful technique we'll explain in the next chapter.

People aren't always aware of the exact triggers that have come to elicit their fear. And as you now know, triggers aren't necessarily logical. Nevertheless, the amygdala is very responsive to them. In order to effectively reduce your anxiety reactions, you need to identify the triggers that provoke your anxiety and then use the approach in chapter 8 to change your amygdala's response to them.

Exercise: Identifying Your Triggers

Take a moment to consider the types of situations in which you experience anxiety. If you're thorough, you may come up with a large number of situations. Don't be discouraged. Instead, consider the big picture. Even though you may think the process of examining so many situations will be overwhelming, you'll probably discover that a lesser number of common triggers lie hidden within this wide variety of situations. For example, you may identify a large number of situations at work that trigger your anxiety but discover a common factor among them when you look more closely, since the same trigger may occur in different situations. Perhaps it's the presence of your boss, the sound of people raising their voices, or situations in which you need to speak before a group. To best identify your triggers, including those that are common to a number of situations, try to consider as many situations as possible in which you feel troublesome anxiety.

When you identify situations in which you experience anxiety, don't forget to consider internal sensations that you may be reacting to. For example, if a pounding heart, dizziness, or the feeling of having to use the bathroom causes you to feel panicky, include those in your list, since internal sensations can also be triggers for anxiety.

You'll find a downloadable Anxiety-Provoking Situations Worksheet that you can use to record these situations at http://www.newharbinger.com/31137 (see the back of the book for information on how to access it). Alternatively, you can create a similar form on a separate sheet of paper using four columns: "Situation that causes anxiety" on the left, then "Level of anxiety," then "Frequency," and finally "Triggers in the situation" on the right. For level of anxiety, rate the intensity using a scale of 1 to 100, where 1 is minimal and 100 is intolerable.

Here's an example to give you an idea of how to use the worksheet: When Manuel used it, the things he listed for "Situation that causes anxiety" were his annual review with his boss, presentations during staff meetings, and arguments with his wife. For the first item, his annual review, he rated his anxiety level at 70 in the second column, and in the third column, reported it happens once a year. In the right-hand column, he identified a number of triggers: the review form he must complete, e-mail reminders from his boss about scheduling the meeting, being in his boss's office, talking to his boss about his performance, the frown he often sees on his boss's face, and the tone of voice that his boss uses when irritated.

In the next row, he listed staff meeting presentations in the first column. He rated the intensity of these at 95, indicating that they are almost unbearable, and noted that he has to present them about once a month. For triggers, he identified the meeting room and the dry mouth he gets when talking, as well as his coworkers looking at him, their criticisms of his ideas, and the facial expressions they make. Then, as Manuel began to write "Arguments with my wife" in the third row, he recognized a pattern in the triggers in these situations. He saw that having to present himself and his ideas in the face of others' criticisms was the source of a lot of his anxiety, and that negative facial expressions are a repeating trigger.

Using the worksheet to identify specific triggers that elicit anxiety for you is very important. Consider the sounds you hear, what you see, the sensations you feel, and what you smell or taste. Also consider what you think or imagine. Bear in mind that the amygdala doesn't always process sensations in the detailed way that you're capable of experiencing them, so a general description of triggers is sufficient. After creating the list, note whether specific triggers appear repeatedly or whether you see a pattern across the different situations that provoke anxiety. This will help you identify your own personal triggers for anxiety.

Sometimes the reason a specific trigger provokes anxiety is obvious. For example, the sight of an elevator would clearly create anxiety in someone with claustrophobia. At other times, the reason for the connection between the trigger and anxiety is less clear, as with Don, the Vietnam veteran who finally figured out that the smell of a certain brand of soap was a trigger. His amygdala clearly recognized an association between the smell of the soap and the danger of combat. While this isn't necessarily a logical association, you can see how it arose. In some cases, the reason a specific trigger elicits anxiety may remain unclear. Fortunately, it isn't necessary to know exactly how the trigger came to cause fearful responding. Regardless of the reason, you can retrain your amygdala, even when you don't know what created the emotional memory.

As you are filling out your worksheet and identifying your own specific triggers for anxiety, you may experience noticeable levels of anxiety due to simply thinking about triggers. As mentioned, the amygdala reacts to triggers in a rather general way. Once the sound of a certain dog growling elicits fear, the sound of other dogs growling is also likely to elicit fear due to *generalization*. This means that even a sound that's similar to a dog growling may result in a feeling of fear. Perhaps most surprising is that simply imagining the sound of a dog growling can be enough to activate the amygdala. This is because when you imagine the sound, you're activating the memory of the sound, and that memory can set off a reaction in the amygdala.

If you feel anxiety as you review the situations you've listed, don't worry about it. Instead, use your emotional responses as an indicator. These emotional responses can help you identify which triggers produce anxiety, allowing you to learn what sets off your amygdala. If you experience some distress, don't be discouraged. In fact, thinking about fear-eliciting triggers is the first step in activating new neural connections and beginning to rewire your brain. So if you begin to feel anxious, tell yourself that you're just heating up the circuits that you need to modify. Take a deep breath and stick with it!

Of course, that may be easier said than done. This work, by definition, is anxiety provoking. You may find the process of thinking about triggers overwhelming. If so, you might wish to undertake this exploration with a therapist, who can support and guide you through the process. Cognitive behavioral therapists are most experienced with this approach, including exposure therapy, which we explain in chapter 8.

Where to Begin

In the next chapter, we'll guide you through the process of retraining your amygdala's reactions to specific triggers. Here, we want to emphasize that ridding yourself of all of your fears is neither possible nor necessary. In fact, it wouldn't be a good idea to eliminate all fears. Fear is appropriate in many situations, such as when you're crossing a busy highway or playing golf as a lightning storm begins. And as mentioned earlier, many fears don't present much of a problem. For example, a fear of flying may have little effect on people who can easily avoid air travel with little consequence. The goal is to begin to modify anxiety reactions that interfere with your ability to live your life in the way you wish. There are three considerations in regard to prioritizing which situations and triggers to work on: the extent to which they interfere with your goals in life, the amount of distress they cause, and the frequency with which they occur. Of course, this isn't an either-or situation. You can choose your focus based on any or all of these factors, but considering them does help you prioritize what to work on first.

Triggers That Interfere with Your Goals in Life

At the end of the introduction, we asked you to consider what your life would be like if your anxiety weren't a limiting factor. Revisiting your thoughts about your goals and hopes is an important step in deciding which triggers to focus on. We highly recommend prioritizing the situations you'll work on by starting with those that most frequently or severely limit your ability to accomplish your daily goals. Which triggers, along with the accompanying emotional responses, are interfering the most severely or frequently and preventing you from living the life you want to live?

Consider Jasmine, who avoided any situation involving public speaking until she enrolled in a nursing program that required her to take a course in public speaking. She quickly saw that her anxiety about public speaking would stand in the way of her goal. This motivated her to seek assistance in reducing her fear of public speaking, and she was soon successful in changing a fear that she'd lived with for years. We encourage you to focus on reducing anxiety in situations where it's an obstacle to achieving your goals. Our intention is to make your goals, not your anxiety, the driving force in your life.

Triggers That Cause Extreme Distress

A second consideration in prioritizing situations and triggers to work on is the level of anxiety you feel in different situations. This is why we ask you to rate intensity in the Anxiety-Provoking Situations Worksheet. If certain situations produce very high levels of anxiety, you may want to focus on them, since they create intense and possibly debilitating stress. Changing how you feel in these situations may give you the most relief.

For example, after two tours of duty in Afghanistan, Verge had strong fear reactions to a variety of sounds, including helicopters, sirens, gunshots, and explosions. But it was explosions that caused the most intense fear, which he rated as over 100. He said fireworks were terrifying, so the Fourth of July and New Year's Eve were a nightmare of repeated panic attacks for him. Verge chose to focus on overcoming his intense fear of explosions first so that he could enjoy these holidays with his family.

Triggers That Arise Frequently

Another consideration is how frequently you find yourself in particular anxiety-provoking situations. Completing the Anxiety-Provoking Situations Worksheet will help you identify the situations that cause anxiety most often. Reducing the anxiety you feel in frequently occurring situations can greatly improve your quality of life because these situations have a greater impact on your daily life. For example, a postal carrier who works in a residential neighborhood and has a fear of dogs may very well want to choose to work on that fear first, since that trigger shows up multiple times on every single workday.

Summary

As you can see, the Anxiety-Provoking Situations Worksheet is extremely useful for identifying the situations you'd like to target for change. Identifying your triggers in these situations will help you know what you need to teach your amygdala. It isn't necessary to change your reactions to all your triggers. Rather, choose to target triggers in those situations in which your anxiety stands in the way of your personal goals and dreams, those that result in the greatest distress, or those you encounter most frequently. In general, the best way to start is with a situation in which reducing your anxiety would significantly improve your life. In the next chapter, we'll show you how to rewire your amygdala to accomplish this.

CHAPTER 8

Teaching Your Amygdala Through Experience

In the previous chapter, we discussed how the amygdala learns to react to certain triggers with fear or anxiety. Once such a reaction has formed, it's difficult to change the pattern and get the amygdala to stop reacting to the trigger. Although you can't easily erase the emotional memory formed by the amygdala, you can develop new connections in the amygdala that compete with those that lead to fear and anxiety. To get the amygdala to create these new connections, you need to expose it to situations that contradict the association between a trigger and negative event. If you show the amygdala new information that's inconsistent with what it's previously experienced, it will make new connections in response to this information and learn from the new experience.

Exposing your amygdala to new information allows you to rewire it in a way that gives you more control over your anxiety. It's similar to adding a *bypass* of a well-traveled area on a highway. When you create a new neural path and practice traveling it again and again, you establish an alternate route that avoids trouble. Responding with fear and anxiety is no longer your only option. You can establish other, calmer responses as a way around your anxiety.

Studies have shown that new learning in the amygdala occurs in the lateral nucleus (Phelps et al. 2004), so you need to communicate new information to the lateral nucleus if you want to train your amygdala to respond differently. Within the brain, there are relatively few connections from the cortex to the amygdala, and the connections that exist don't communicate directly with the lateral nucleus or the central nucleus. The connections from the cortex seem to send their messages to *intercalated neurons,* a collection of nerve cells that lie between the lateral nucleus and the central nucleus. Though these neurons allow the cortex some influence over ongoing responses, the cortex doesn't seem to have direct connections to the lateral nucleus.

You need to specifically retrain the amygdala if you want to reduce its influence on your brain, emotions, and behavior. By practicing the exposure techniques described in this chapter, you can communicate new information to the lateral nucleus and rewire the pathways associated with specific triggers.

If you think about it, you're surrounded by examples of people who have overcome even what are thought to be naturally inborn fears. For example, in big cities, you can see people suspended from ropes cleaning the windows of tall skyscrapers, apparently quite calm as they go about their day's work. People who participate in water skiing, horseback riding, or ballroom dancing may have had to overcome fears to participate in these activities. Learning to swim or drive often requires people to overcome anxiety.

Repeated exposure to a seemingly threatening situation without anything negative occurring can teach the amygdala that the situation doesn't require a fearful response. You can overcome fears if you give your amygdala experiences that teach it to feel safe in related situations. This is the power of exposure.

Exposure-Based Treatment

Among the various types of therapy for anxiety-based difficulties, especially panic attacks, phobias, and obsessive-compulsive disorder, none has been as dramatically successful as *exposure therapy* (Wolitzky-Taylor et al. 2008). In this approach, people are exposed to situations or objects they fear, sometimes in a gradual way and sometimes more abruptly. During each exposure, anxiety typically rises, often to an uncomfortable level, and then begins to subside. The key is to let the anxiety response run its course, peaking and then lowering, without escaping the situation. In this way, the amygdala begins to pair a previously feared situation with safety.

The power of exposure therapy lies in giving the amygdala new experiences that prompt it to make new connections. According to psychologist Edna Foa—who has conducted extensive studies of exposure—and colleagues, its effectiveness comes from the *corrective information* it provides (Foa, Huppert, and Cahill 2006). The learning experiences offered by exposure show the amygdala that triggers that previously evoked fear and anxiety are actually quite safe. Exposure therapy is a highly effective way to speak the language of the amygdala.

Systematic desensitization and flooding are two examples of exposure-based treatment. *Systematic desensitization* involves learning relaxation strategies and approaching feared objects or situations in a gradual manner. Typically, this occurs in a slow but steady process, gradually working through situations that elicit increasingly more anxiety as the therapy proceeds. With *flooding*, in contrast, people jump right in with the most fear-provoking situation, and the exposure may last for hours. Flooding is a more intense approach, but it also provides relief from anxiety much more quickly.

With either approach, in most cases people initially confront feared situations mentally, by imagining themselves in the feared situation. But ultimately, they must directly experience the situation, usually repeatedly. Obviously, this is a challenging form of treatment, but research shows that it's exactly the approach needed to rewire the amygdala (Amano, Unal, and Paré 2010). Therefore, the more you practice exposure, the more likely it is that your amygdala will respond calmly to previously feared situations and triggers.

You may wonder whether the gradual approach of desensitization or the more rapid approach of flooding is more effective. Research indicates that intense, extended exposure to triggers that produce fear (flooding) is more rapid and effective than a gradual approach (Cain, Blouin, and Barad 2003). But for an approach to be effective, a person must be willing to use it. Not surprisingly, anxious people tend to be more likely to try a gradual approach like desensitization, rather than flooding. In the end, either approach works, because ultimately both allow the amygdala to experience previously feared stimuli without any negative results.

Because exposure-based treatments are so effective, they are one of the most frequently recommended approaches to reducing anxiety. Many people who have learned to cope with their anxiety have had personal experiences or professional treatment that involved exposure. If you haven't engaged in exposure-based treatment, we recommend that you seek a professional to guide you through the process, as evidence indicates that the support of a therapist is very helpful. If you've already had exposure therapy, we hope that this book will help you understand why exposure works. And if you've tried exposure therapy and it wasn't effective or lasting, we hope this book will help you understand why. If you give it a try again and follow the approach outlined in this chapter, we believe it will be helpful.

Of course, exposure therapy isn't easy. By definition, it produces anxiety because it involves deliberately engaging in anxiety-provoking experiences. Knowing that this process is required to rewire your brain will help you rise to the challenges and make the stress of the experience more tolerable.

Nothing speaks to the amygdala more effectively than experiences that activate the neurons associated with feared situations and objects. Your amygdala is constantly monitoring your experiences and creating connections between neurons that indicate what it believes to be safe versus dangerous. Exposure-based treatment gives the amygdala opportunities to make new connections and practice those connections over and over again.

The Basics of Rewiring: Activate to Generate!

The amygdala must have particular experiences for rewiring to occur. During exposure, you need to experience the sights, sounds, and other stimuli that create anxiety in order to activate the exact neural circuitry that holds the emotional memories you want to modify. Activating these circuits creates the potential to develop new connections between different neurons—connections that will modify the amygdala's responses. Again, you must *activate* the neurons to *generate* these connections. You must experience fear or anxiety in order to conquer it. There's more than cowboy wisdom in the old adage "You gotta get back on the horse that threw you."

When people are left to their own devices, the amygdala typically doesn't get the learning experiences it needs to change its responses to feared situations. In fact, the anxiety response often prevents effective exposure from occurring. Consider a grandmother with a fear of flying who receives the gift of an airline ticket for a trip to visit her extended family thousands of miles away. As she contemplates packing for the trip or arrives at the airport to board the plane, her anxiety increases, creating excellent opportunities for exposure. But she doesn't realize that her anxiety means she's in the best position to rewire the activated circuits and change her amygdala's response to the situation. Instead, her most natural reaction to her anxiety in such situations will probably be to try to avoid taking the trip. You can reason with her that flying is safer than driving, and she may understand that or even reason with herself. But her amygdala isn't operating on the basis of reason; it's simply activating established connections that produce a stress response.

When facing feared situations, the discomfort can sometimes seem unbearable, and the desire to escape, irresistible. Yet if that grandmother avoids taking the flight, she will miss an opportunity for exposure, and miss an opportunity to spend time with her family. This dynamic of experiencing anxiety and then escaping it by avoiding the situation just serves to maintain anxiety, and this is exactly what makes anxiety reactions so difficult to modify. In this way, anxiety can be self-perpetuating.

To remind yourself why it's necessary to experience anxiety, remember the phrase "activate to generate." This is what's required for learning to occur in the amygdala. The activation of neurons underlies the effectiveness of exposure-based therapy. If you want to generate new connections, you must activate the circuits that store the memory of the feared object or situation (Foa, Huppert, and Cahill 2006). The emotional arousal and anxiety that occurs is a sign that you're activating the right circuits. In fact, evidence shows that people who have higher levels of emotional arousal benefit most from exposure during initial exposure experiences (Cahill, Franklin, and Feeny 2006). This may also explain why flooding works more rapidly than systematic desensitization.

Animal research and brain imaging indicate that exposure, the process of experiencing a situation or object that causes anxiety while nothing negative happens, allows another portion of the brain to exert some control over how the amygdala responds (Barad and Saxena 2005; Delgado et al. 2008). This other portion of the brain lies in the frontal lobes, and research on humans shows that an area called the ventral medial prefrontal cortex seems to be involved (Delgado et al. 2008). During exposure, learning occurs in the amygdala, and the memory of this learning is stored by the ventral medial prefrontal cortex. The fear learned in and stored by the amygdala isn't erased (Phelps 2009), but other circuitry is developed, and new, calmer responses are learned.

An analogy may help you remember how important it is to activate the circuitry of anxiety, even though this is an uncomfortable experience. When you're making a cup of tea, you'll have better results if the water is hot. Putting tea leaves or a tea bag in a cup of cold water won't be as effective in allowing the flavor of the tea to be infused into the water. In a similar way, your neural circuits need to be activated (or hot) to make new connections. When it comes to anxiety, you need to be exposed to the heat if you want to rewire your neural circuitry.

Understanding Exposure Through Practice in Diagramming

Consider the unfortunate experience of a young boy who was scratched by a cat. The cat, a neutral object that became a trigger, is associated with the scratch, a negative event that caused pain. As a result, cats have come to elicit anxiety. Afterward, when the boy sees a cat, he experiences anxiety and isn't at all interested in playing with it.

If we want to help him create new circuitry and change his fear of cats, we need to expose him to friendly cats in order to retrain his amygdala. When he sees or touches a cat under positive circumstances (while petting it and enjoying its softness, being amused by its antics, and so on), his amygdala can be stimulated to establish new circuitry related to cats. The more the child observes or interacts with cats in the absence of negative events, the stronger the new neutral or positive connections will become and the less anxiety he'll experience. With repeated exposure to friendly cats, the child's amygdala will create a bypass around his fear and anxiety.

Of course, during exposure the child is likely to feel and express fear of the cat. But this exposure is necessary to activate the neurons we wish to rewire. There's no way to change the circuitry the lateral nucleus has created without giving the amygdala new experiences with a cat and, as a consequence, creating some anxiety. In fact, the child's anxiety is a good indication that the correct circuits in the amygdala have been activated and are ready for new learning.

To diagram the process of creating new connections, we can build on the basic diagram we've been using (see figure 7). This time, we link the cat not with a scratch but with a positive experience, such as watching a playful cat chasing a string or petting a purring cat. In this way, cats will come to elicit more positive feelings, perhaps calmness or pleasure. This new connection can compete with the previous connection between cats and anxiety, providing a route around the anxiety response. The more the child is exposed to positive experiences with cats, the stronger the path will become and the more likely it is that the child will feel positive emotions, rather than anxiety, when he encounters cats in the future. Repeated exposure creates this new, alternative response.

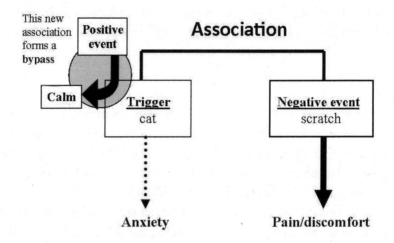

Figure 7. Creating a new neural connection.

Taking the Plunge

Exposure is a "no pain, no gain" situation. You must expose your-self to feared situations and allow yourself to experience anxiety if you want to change your response. The optimal condition for learn-ing to take place in the amygdala is when the neurons are excited, just as the optimal condition for building muscle mass in the body is when the muscle fibers are fatigued. In a parallel way, as you do more repetitions, you grow stronger. You can think of exposure as a way of providing exercises that will train your amygdala.

We assure you that a great deal of evidence indicates that expo-sure is a highly effective way of changing the connections in the brain that are responsible for anxiety. Still, it's difficult to deliber-ately put yourself in a situation designed by its very nature to distress you, and sometimes it's downright impossible. You shouldn't try exposure until you're confident you'll follow through, because it's possible to actually strengthen anxiety if you leave the exposure situ-ation before your anxiety decreases.

Because exposure that's done incorrectly can strengthen anxiety, we recommend working with a therapist who's experienced with exposure. This will ensure that you get the best treatment. You also should carefully choose when to use exposure and when not to, so you can utilize this powerful tool to help you gain control over the most important aspects of your life. Use exposure with situations that will have the most impact on your life, and don't put yourself through exposure when changing your fear response isn't necessary. For example, if you don't need to get over a fear of snakes, don't work on that!

Exposure won't be terribly distressing in every moment, especially if you choose gradual approaches. And when exposure is quite challenging, you can strengthen your resolve by reminding yourself that you'll experience changes in your anxiety relatively quickly. A good analogy for illustrating the transformative power of exposure is going for a swim. Have you ever dipped your toes into a pool or lake and winced at the cool temperature of the water? As you wade further in, you're aware of the coolness of the water as it gradually reaches your stomach and chest. After a period of time, however, your body adjusts, and you find yourself comfortable in the water. You smile at others who are in just up to their knees and complaining that it's cold. The same process of adjustment occurs with exposure. Your amygdala will adapt if you remain in the situation. When you're practicing exposure exercises and you feel your anxiety decrease, you'll know that you've gotten the amygdala's attention and are making progress!

Medication Considerations

If you're on antianxiety medication, be aware that some medications can assist you in the exposure process, whereas others make it harder for your amygdala to learn. Benzodiazepines, such as Valium (diazepam), Xanax (alprazolam), Ativan (lorazepam), and Klonopin (clonazepam), may interfere with exposure. These drugs have a tranquilizing effect on the amygdala, which helps keep anxiety in check. However, the process of rewiring is based on activating the amygdala and creating anxiety to generate new learning. New learning is less likely to occur in a brain medicated with benzodiazepines. In fact, research has shown that taking benzodiazepines decreases the effectiveness of exposure-based treatment (Addis et al. 2006); and multiple studies have found that the people who benefit most from exposure-based therapy aren't taking them (for example, Ahmed, Westra, and Stewart 2008).

On the other hand, certain medications assist with the process of exposure, including selective serotonin reuptake inhibitors (SSRIs) and serotonin-norepinephrine reuptake inhibitors (SNRIs). The SSRIs include medications such as Zoloft (sertraline), Prozac (fluoxetine), Celexa (citalopram), Lexapro (escitalopram), and Paxil (paroxetine). SNRIs include medications such as Effexor (venlafaxine), Pristiq (desvenlafaxine), and Cymbalta (duloxetine). Research indicates that SSRIs and SNRIs promote growth and change in neurons (Molendijk et al. 2011). Therefore, these medications may make it more likely that brain circuitry can be modified by experience.

Of course, it's important that you work with your health care providers when making any adjustments to medications. If you'd like to learn more about various antianxiety medications and when they may or may not be helpful, you can download a bonus chapter on this topic, "Medications and Your Anxious Brain," at http://www.newharbinger.com/31137. (See the back of this book for information on how to access it.)

Strengthening New Connections

To create new connections most effectively, you must engage in multiple exposures to the triggers that cause your anxiety. Remember, you have to activate the fear circuitry to generate new connections. Repeated exposure will not only form these new connections but also strengthen the new circuitry so it can override the fear circuitry previously established by the lateral nucleus. So, for example, if you're trying to get over a fear of elevators, the most effective approach is to ride a variety of elevators in different settings.

Of course, it's essential that your experiences during exposures be neutral or positive. Continuing with the previous example, make sure your exposures to riding in elevators are uneventful and calm. This doesn't mean the exposures will be anxiety free, obviously. Remember, courage is not the absence of fear; courage is acting *despite* fear. The more you experience anxiety and stay in the situation long enough for your fear to diminish, the stronger the new circuitry will become.

Designing Exposure Exercises

In chapter 7, you learned how to complete the Anxiety-Provoking Situations Worksheet. Select one situation from your worksheet to begin with, keeping in mind the considerations regarding prioritization discussed at the end of chapter 7 (choosing a situation that keeps you from achieving your goals, a situation that creates a great deal of distress, or a situation that comes up often). Start by reviewing the anxiety-producing triggers in this situation. Again, we recommend working with a doctor or therapist who understands exposure-based treatment and can provide support and guidance.

Once you've selected the situation you want to focus on, decide whether you prefer the slow, systematic desensitization approach or want to plunge right in with flooding. In the systematic desensitization approach, you'll take the process step-by-step in a gradual way, working your way up to the most challenging situations over time. In the flooding approach, you'll begin with some of the most challenging situations and work through them in an intense process. Again, flooding is quicker, but either approach will work. In this chapter, we'll guide you through the systematic desensitization approach, helping you break down the process into a hierarchy of steps. However, you can easily use flooding by simply starting with some of the most difficult situations.

Creating a Hierarchy for Exposure

An *exposure hierarchy* is a rank-ordered list of steps that you'll face sequentially to learn new responses to a specific situation. In a hierarchy, you break down a specific anxiety-provoking situation into smaller components and begin by facing those that are least anxiety provoking, eventually progressing to those that are more challenging.

A woman who fears shopping in a mall will serve as an example. To help her construct her hierarchy, we'd begin by asking her to identify the most stressful behavior that could be required of her. Let's say she responds with "Going into a crowded store and standing in line until I make a purchase." Then we'd ask her to identify a related behavior that would elicit some anxiety but that she's fairly confident she could carry out. To that, she might respond, "I could drive to the parking lot and find a parking space." To construct her hierarchy, we'll use these as the extremes and fill in intermediate steps between them. We'd then ask the woman to come up with at least five more related behaviors that would provoke anxiety at levels between the two extremes. Her list might look something like this:

- *Selecting an item to purchase*

- *Holding an item and thinking about purchasing it*

- *Walking from the car to the mall entrance*

- *Asking a clerk a question about an item*

- *Walking around the mall with a supportive friend*

- *Feeling nausea (due to anxiety) in a public place*

- *Walking around the mall alone*

- *Walking around the mall alone when it's crowded*

Next, we ask her to order these behaviors from least to most anxiety provoking, placing them between the first two extremes she identified. A scale of level of anxiety ranging from 1 to 100 is useful for putting items in rank order so that the level of anxiety increases with each step. Sometimes what varies is what the person has to do. For example, this anxious shopper feels more anxiety if she has to purchase something than if she simply has to walk around in the mall. In other cases, the triggers might differ, such as being in a crowd or asking a salesperson a question. Other aspects of the situation that might differ include whether another supportive person is present or physical proximity to a trigger. You would put your own hierarchy steps in order by considering the level of anxiety you'd be likely to experience in each step. Then it's time to start practicing, working from the least anxiety-provoking item to the one that would create the most anxiety.

Here's an exposure hierarchy for the anxious shopper. Notice how the situations are ordered into steps in terms of increasing anxiety. As you'll see, there's a large increase in anxiety rating between step 4 and step 5.

Step number	Description of behavior or situation	Level of anxiety (1–100)
1	Driving to the parking lot and finding a spot	15
2	Walking from the car to the mall entrance	15
3	Walking around the mall with a supportive friend	20
4	Walking around the mall alone	30
5	Feeling nausea in a public place	50
6	Walking around the mall alone when it's crowded	60
7	Selecting an item to purchase	70
8	Holding an item and thinking about purchasing it	75
9	Asking a clerk a question about an item	80
10	Standing in line until I make a purchase	90

Exposure isn't easy. Again, if possible, find a therapist who specializes in exposure therapy to guide and encourage you through the process. In addition to helping you work through a hierarchy, your therapist may have you do exercises that can help desensitize you to the physical sensations resulting from anxiety, such as heart palpitations, shallow breathing, and light-headedness. These might include *interoceptive exposure*, which uses simulations such as vigorous activity, intentional hyperventilation, breathing through a straw, or spinning in a chair to help people become more accustomed to some of the physical symptoms of anxiety.

If you have obsessive-compulsive disorder, an exposure hierarchy can also help you learn to resist compulsions. You simply create a similar hierarchy of situations that provoke compulsive behaviors and then expose yourself to those situations without allowing yourself to perform compulsions in response. So if touching canned goods results in a compulsion to wash your hands, you'd touch canned goods repeatedly without washing your hands. This process is called *exposure with response prevention*.

Practicing Exposures

Once you've created a hierarchy, the goal is to eventually accomplish each step, staying in the situation until your anxiety diminishes or the compulsion is reduced. We recommend using deep breathing and other relaxation techniques from chapter 6 to cope with the anxiety you feel during each session. You need not experience a high level of anxiety for rewiring of the circuits to occur, but if anxiety is high during exposure, this can speed the process of change (Cahill, Franklin, and Feeny 2006).

During each exposure, it's vital not to leave the situation in fear, as this will strengthen the fear circuitry. You need to stay in the situation until you feel your anxiety decrease, preferably by half. In other words, if you rate your initial anxiety in a situation at 80 on a scale of 1 to 100, don't leave the situation until your anxiety diminishes to 40 or below. (Often you can actually feel when the amygdala registers the new information and calms down.) Your amygdala needs to learn that it's safe and that escape isn't necessary. Remember, this is something you must *show* the amygdala; it only learns through experience.

Exposure to each step must be done repeatedly for change to occur in your amygdala. Typically each repetition of a specific step is easier than the previous one, but sometimes there are ups and downs. Once you've surmounted the most difficult item on your hierarchy and have accomplished your goal, you can choose another feared situation to work on and approach it in the same way.

The more you're limited by your anxiety, the more frequently you'll need to practice exposure in order to regain control of your life. Also, make sure you plan in advance. If you don't schedule your exposures and plan to repeat them, you won't rewire your brain and reduce your anxiety. Finally, we recommend that you reward yourself for your progress each time you surmount a step. You deserve a reward for putting yourself through these difficult exercises!

In every exposure session at each step along the way, carefully monitor your thoughts so that your cortex doesn't unnecessarily increase your anxiety by engaging in self-defeating or anxiety-provoking thoughts. You're trying to reduce amygdala-based anxiety, not worsen it with thoughts from the cortex. Stay focused on the step that you're facing, and don't anticipate other situations that are higher on your hierarchy.

Helpful Hints

When practicing exposure, there are a few things you should refrain from doing. As a reminder, don't leave the situation while your fear is still high. If you flee and then feel relief, you'll teach your amygdala that escape is the answer. This will only increase your anxiety in the future, as your amygdala tries to compel you into escaping again, so resist the urge to flee. Stay in control of your behavior; don't let anxiety take control of you.

As mentioned, it's also important to monitor your cortex for thoughts that may increase your fear. The cortex is capable of making the situation worse by thinking negative thoughts. When you detect self-defeating or anxiety-provoking thoughts, try to substitute helpful coping thoughts, such as these:

- "I expect my fear to rise, but I can manage it."

- "Stay focused on this situation. This is all I have to manage."

- "Keep breathing. This won't last long."

- "Relax my muscles. Let the tension go."

- "I'm activating my fear circuits to change them. I'm taking control."

- "Just stay until the fear decreases. It will decrease if I wait."

- "I must activate to generate."

Finally, don't use safety-seeking behaviors, which can undermine all of your hard work during exposure. Here are some examples of safety-seeking behaviors to avoid:

- Having extra medicine available so you can use it in an emergency

- Having a safe person present for all of your steps

- Carrying lucky charms of various kinds

- Holding on to objects

- Wearing sunglasses

- Sitting in a particular position or location

- Talking on a cell phone

- Staying near an exit or a bathroom

When you use safety-seeking behaviors, exposure is only partial and doesn't result in the changes in the brain you're seeking. If you do use safety-seeking behavior during some of your steps, make sure to eliminate it in later steps to ensure that all of your hard work during exposure has the desired effect.

Summary

In this chapter, you've learned how to rewire your amygdala by activating it in the presence of triggers. You've learned how to use a hierarchy to expose your amygdala to triggers in a gradual manner. The most important element of exposure therapy is practice, practice, practice. The only way the amygdala learns is through experience. At times it's going to be upsetting, even daunting. But if you really want to overcome your anxiety, you need to do this difficult work. Remember, it's a "no pain, no gain" proposition. Just like having toned abdominal muscles requires doing a lot of sit-ups, changing fear responses requires that you face feared situations and conquer them one step at a time. Building a bypass and using it often is the best means of achieving lasting relief from anxiety. Your amygdala can and will change if you're willing to put some time, effort, and courage into challenging your fears and teaching your amygdala new responses.

CHAPTER 9

Exercise and Sleep
Tips for Calming
Amygdala-Based Anxiety

A variety of neuroimaging studies and neurophysiological experiments have shown that the amygdala can be strongly influenced by both exercise and sleep. Exercise has surprisingly powerful effects on the amygdala, surpassing many antianxiety medications in effectiveness. Sleep also has a strong impact on the amygdala's functioning, with lack of sleep leading to heightened anxiety. In this chapter, you'll learn how to make specific changes in your lifestyle that can ease amygdala-based anxiety and also reduce your stress level and improve your psychological health more generally.

Using Exercise to Cope with Anxiety

The fight, flight, or freeze response is programmed into the amygdala. Instead of fighting this ancient response, perhaps we should try to work with it at times. If your sympathetic nervous system is activated, you can put it to use as nature intended. Instead of resisting your body's preparations to fight or flee, why not look for opportunities to work with that instinct and utilize your muscles in ways that will decrease the amygdala's activation?

Brief periods of aerobic exercise can be very effective in reducing muscle tension. And as you learned in chapter 6, relaxing your muscles can help ease anxiety. If you run or walk briskly when you feel anxious, you'll make use of muscles that have been prepared for action. This will lower levels of adrenaline and use up glucose released into the bloodstream by the stress response. And after you exercise, you'll experience substantial, long-lasting muscle relaxation. In the sections that follow, we'll examine some of the effects of exercise on the body and brain to help illuminate why exercise is such a useful strategy in coping with anxiety.

Effects of Exercise on the Body

The type of exercise that will be most helpful for easing the reactions of the SNS is aerobic exercise, which makes use of large muscle groups in rhythmic movements at a moderate level of intensity. Common forms of aerobic exercise include running, walking, cycling, swimming, and even dancing.

In addition, adhering to a regular exercise program can reduce SNS activation more generally (Rimmele et al. 2007), including decreasing its impact on blood pressure (Fagard 2006) and heart rate (Shiotani et al. 2009). This helps counter the symptoms of an activated amygdala. Of course, exercise has many other benefits for the body. For example, aerobic exercise tends to increase a person's metabolic rate and energy level. So if you use exercise to help you cope with anxiety, you'll get a windfall of extra benefits.

If you haven't been exercising regularly, please do consider potential risks. Consult your doctor before you begin, and increase your activity level gradually, not all at once. Bear in mind that some forms of exercise, such as jogging, are high-impact activities that can lead to a variety of injuries. However, don't let a lack of experience discourage you, because almost anyone can do simple forms of exercise, such as walking, without much difficulty or risk.

Exercise and Anxiety

We highly recommend exercise as a strategy for reducing anxiety because, quite simply, it works. A variety of studies have demonstrated that aerobic exercise can ease anxiety (Conn 2010; DeBoer et al. 2012). Reductions in anxiety are measurable after only twenty minutes of exercise (Johnsgard 2004). That's less time than it takes for most medications to begin working. The reduction in anxiety is greatest for people who have higher levels of anxiety to begin with (Hale and Raglin 2002). Furthermore, exercise is helpful for people who are sensitive to the symptoms of anxiety, such as increased heart rate or breathlessness, because those sensations are also associated with exercise. Therefore, exercise can serve as a form of exposure that decreases people's discomfort about those sensations (Broman-Fulks and Storey 2008).

Generally, exercise results in decreased muscle tension for at least an hour and a half afterward, and reductions in anxiety last from four to six hours (Crocker and Grozelle 1991). If you consider that twenty minutes of sustained exercise may result in hours of relief from tension and anxiety, the benefits are clear. In fact, if you anticipate that a particular event or phase of your day may amp up your anxiety, a carefully timed exercise routine may allow you to get through it with less anxiety. In other words, you may be able to achieve a tranquilizing effect without taking tranquilizers.

Consider Alli, a seventeen-year-old who was anxious about an upcoming family reunion in her family's home. Her difficulties with social anxiety made the event seem like a nightmare for her, and she feared feeling trapped. When her therapist suggested that she go for a run if she began to feel panicky during the reunion, Alli literally rolled her eyes. But on the day of the reunion she tried it, in her words, "Mostly because I just wanted to get out of there!" After a brief run around the neighborhood, she came back into the house with a feeling of relief that surprised her. She was able to talk to her aunts and uncles without anxiety and later said, "I really believe that my amygdala thought I had escaped from the danger, and it calmed down!" She was sold on the anxiety-reducing benefits of exercise from that day forward.

Exercise doesn't just reduce anxiety in the moment or for a few hours afterward. Research shows that following a regular exercise program for at least ten weeks can reduce people's general level of anxiety (Petruzzello et al. 1991).

Effects of Exercise on the Brain

The finding that exercise reduces anxiety has led to research into what's happening in the brain to account for this. You're probably familiar with the runner's high, in which people feel a sense of euphoria after crossing a certain threshold of exertion. Extended or intense aerobic workouts have been shown to cause the release of endorphins into the bloodstream, and these neurotransmitters have been proposed as the cause of that feeling of exhilaration (Anderson and Shivakumar 2013). "Endorphin" is a shortened name for "endogenous morphine," meaning "morphine-like substances produced naturally in the body." And as that suggests, these compounds can reduce pain and produce a sense of well-being through their effects on the brain.

Animal studies have helped illuminate what may be happening in the brain after exercise. When laboratory rats are offered free access to a running wheel, they generally make use of it. What's more, the level of endorphins in their brains increases and remains elevated for many hours afterward, only returning to typical levels after about ninety-six hours (Hoffmann 1997). This finding indicates, once again, that the effects of exercise on the brain last much longer than duration of the exercise period and may in fact persist for days. It's quite possible that when you exercise, you're raising your endorphin level not just for that day, but for days afterward.

Effects of Exercise on the Amygdala

Additional research involving rats running on their wheels has shown that exercise changes the chemistry of the amygdala, including altered levels of the neurotransmitters norepinephrine (Dunn et al. 1996) and serotonin (Bequet et al. 2001). Exercise appears to affect a certain type of serotonin receptor that's found in large numbers in the lateral nucleus of the amygdala (Greenwood et al. 2012). Regular exercise seems to make these receptors less active, resulting in a calmer amygdala that's less likely to create an anxiety response (Heisler et al. 2007). This calming effect on the amygdala after regular exercise has been found in humans (Broocks et al. 2001), as well as in mice and rats.

Effects of Exercise on Other Parts of the Brain

Scientists were surprised when they first discovered that exercise could promote brain cell growth in rodents. Twenty years ago, new cell growth in the brain wasn't considered possible. Now researchers know that regularly running on wheels increases the levels of certain neurotransmitters and promotes new cell growth in rats (DeBoer et al. 2012). Research also confirms that exercise promotes factors that stimulate cell growth in human brains (Schmolesky, Webb, and Hansen 2013), strengthening the evidence for neuroplasticity—the ability of the brain to change. Scientists have learned that simply exercising can increase levels of neurotransmitters and promote the growth of new cells in the human brain.

Exercise produces changes that affect the cortex, as well as the amygdala. Endorphins have effects in the cortex, and changes in the levels of these and other neurotransmitters affect a variety of regions in the brain. Exercise also produces a protein (brain-derived neurotrophic factor) that promotes growth of neurons in the brain, particularly in the cortex and hippocampus (Cotman and Berchtold 2002). In addition, neuroimaging studies of activity in the brain indicate that exercise tends to modify activation of certain areas of the cortex. For example, after running on a treadmill for thirty minutes, men showed greater activation in the left frontal cortex compared to the right frontal area (Petruzzello and Landers 1994). Greater left frontal activation has been associated with a more positive mood, suggesting that exercise can stimulate the cortex in a manner that produces more positive feelings. Those positive feelings are likely to help reduce anxiety.

Considering What Type of Exercise Is Best for You

The best type of exercise for you, both physically and mentally, is exercise that meets the following four criteria:

- You enjoy doing it.

- You'll keep doing it.

- It's moderately intense.

- Your doctor approves it.

This means you should choose one or two types of exercise to engage in at least three times a week for thirty minutes each time. Whatever you choose, remember that getting your heart pumping and your blood flowing has many benefits. Once you feel the improvement in your mood and reductions in your overall stress level, you should find it easier to stick with an exercise program.

Exercise: Assessing Your Exercise Quotient

This brief exercise will help you assess your current exercise patterns and strengthen your commitment to a regular, long-term program of physical activity. Take some time to consider all of the following questions:

- *How often do you exercise each week, and how long does each period of exercise last?*

- *Do you feel less anxious after exercise?*

- *If you don't exercise regularly, would you consider beginning an exercise program to decrease the SNS activation that anxiety creates?*

- *Which type of exercise most appeals to you?*

Sleep: An Active Time for the Brain

Most people know how much more refreshed and alert they feel when they've gotten a good night's sleep, but few truly grasp how important sleep is for the brain. People tend to see sleep as a period in which the brain shuts down, but sleep is actually a very active time for the brain. Just like your heart or immune system, your brain continues to work while you sleep, and in fact, during certain periods of sleep, it's more active than at any time when you're awake (Dement 1992). As you sleep, your brain is busy making sure that hormones are released, needed neurochemicals are produced, and memories are stored.

Yet getting good, restful sleep is often a challenge for people who struggle with anxiety. When anxiety interferes with sleep, it's due to the influence of the amygdala. By promoting activation of the SNS, the amygdala can keep you in an alert state that prevents you from cycling down into deep sleep. Worries produced in the cortex can compound the problem by exposing you to distressing thoughts that contribute to the amygdala's activation of the SNS. Worse, if you don't take steps to ensure that you get good sleep, you run the risk of making your anxiety even worse, since a lack of sleep can make the amygdala prone to more anxious responding.

Sleep Difficulties

If you have difficulty falling asleep or awaken before you need to and can't get back to sleep, it's important for you to read this section. Many people are unaware that sleepless nights have detrimental effects on their health, their brain, and specifically the amygdala. Don't assume that you're getting enough sleep if you don't feel tired. When you're sleep deprived, you can still feel alert or even energetic in stimulating situations. And because anxious people are often in an alert state, with an activated SNS, they may not feel sleepy and therefore assume that they aren't sleep deprived. It may be that they are and just don't recognize it. Be aware that sleep deprivation can show up in many forms, including increased anxiety or irritability, difficulty concentrating, or lack of motivation.

Exercise: Assessing Whether Sleep Difficulties Are an Issue for You

To help you determine whether you have sleep problems, read through the statements below and check any that are true for you.

_____ I'm often restless and find it difficult to fall asleep when I go to bed.

_____ I've used medications or alcohol to help me sleep.

_____ I need complete silence to sleep. Any noise will prevent me from relaxing.

_____ It often takes me more than twenty minutes to fall asleep.

_____ I often feel drowsy, fall asleep, or nap during the day.

_____ I don't go to bed or wake up at a consistent time.

_____ I awaken too early and can't get back to sleep.

_____ I don't sleep soundly. I just can't relax.

_____ When I get out of bed in the morning, I don't feel rested.

_____ I dread trying to go to sleep at night.

_____ I depend on caffeine to get me through the day.

The more of these statements you check, the more likely it is that you have sleep debt. *Sleep debt* occurs when people haven't been getting as much sleep as they need and the hours of missed sleep start accumulating. Most adults need between seven and nine hours of sleep per night. Each night you miss an hour or so of sleep, your debt grows. So even if you get enough sleep on a given night, you may still feel sleepy or irritable the next day as a result of an accumulated sleep debt.

Lack of Sleep and the Amygdala

Poor sleep has detrimental effects on the human brain. People who don't get enough sleep have difficulty concentrating, problems with memory, and poorer health in general. But in this chapter we're particularly interested in how lack of sleep affects the amygdala, so let's take a look at what research into this question reveals. Studies have shown that the amygdala reacts more negatively to a lack of sleep than other parts of the brain.

In one study (Yoo et al. 2007), a group of people was kept from sleeping for one night, and another group was allowed to sleep normally. Then, at about 5 p.m., all were brought into a laboratory and shown a variety of images, both positive and negative, while scientists used functional magnetic resonance imaging to observe how their amygdalas reacted. The sleep-deprived people, who had gone approximately thirty-five hours without sleep, had about 60 percent more amygdala activation in response to the negative images (Yoo et al. 2007). So be aware that if you're going without sleep, you're making it more likely that your amygdala will be reactive and cause you to experience anxiety or other emotional reactions, such as anger and irritability.

When we sleep, we go through different stages of sleep in a particular pattern. We cycle through these different stages in a repetitive manner, with rapid eye movement (REM) sleep typically occurring several times over the course of the night. REM sleep is the stage of sleep during which dreaming occurs. It's also a time when memories are consolidated and neurotransmitters are replenished. Researchers have found that lower reactivity in the amygdala is associated with getting more REM sleep (van der Helm et al. 2011). This suggests that getting good sleep, especially sufficient REM sleep, can help calm the amygdala.

As you work on getting adequate sleep, it's important to understand when REM sleep occurs. REM sleep occurs later in the sleep cycle, and phases of REM sleep become more frequent at the end of the overall sleep period. Many people don't realize that a long period of sleep is necessary for getting into these stages of REM sleep. Therefore, four hours of sleep followed by an hour of wakefulness and then another four hours of sleep isn't equal to eight hours of sleep. When you return to sleep after being awake for even just half an hour, the sleep cycles start over from the beginning, so it will take many more hours to get through an entire sleep period. It isn't like returning to watching a movie where you left off. It's like having to go through the whole movie again from the beginning.

Coping with Sleep Difficulties

After reading this information about sleep, you may be thinking, *I want to get good sleep, but it isn't easy!* Of course, our current twenty-four-hour culture, with media, shopping, and restaurants available at all hours, can keep us from getting sufficient sleep on a regular basis. Certain stages of life also make one vulnerable to sleep deprivation, including the college years or the first months of parenthood.

Many people treat sleep as a luxury that can be disregarded when necessary. To calm anxiety, you need to resist influences that interfere with sleep. However, anxiety itself often impairs people's ability to sleep, with difficulty falling asleep or early awakening both being quite common. When coping with these difficulties, it's useful to know which approaches will help and which will actually worsen the problem.

The best approach to improving sleep is to take a careful look at your sleep-related routines to make sure that they're healthy. The following sleeping practices can really assist you in achieving a good night's sleep:

- Before you go to bed, practice a routine set of relaxing rituals.

- Eliminate light stimulation for at least an hour before bed.

- Exercise during the day.

- Establish a consistent bedtime and waking time.

- Avoid napping.

- Near bedtime, replace activating thoughts with relaxing ones.

- If worries haunt you at bedtime, schedule a worry time during the day.

- Ensure that your sleeping environment is conducive to sleep.

- Avoid caffeine, alcohol, and spicy foods in the late afternoon and evening.

- Use relaxing breathing techniques to prepare for sleep.

- If you can't fall asleep after thirty minutes in bed, get up and do something relaxing.

- Use your bed primarily for sleep.

- Avoid using sleep aids.

The above suggestions are all examples of good *sleep hygiene*. (For a downloadable document with more details on good sleep practices, visit http://www.newharbinger.com/31137; see the back of the book for information on how to access it.)

Summary

Clearly, lifestyle habits can have a strong influence on your amygdala. If you engage in regular aerobic exercise, especially exercise that uses large muscle groups, the positive effects on both your amygdala and your cortex can help improve your mood. Exercise also increases neuroplasticity, making both your amygdala and your cortex more responsive to the rewiring you're attempting to achieve. In addition, ensuring that you're getting sufficient good-quality sleep can calm the amygdala and make it less reactive to whatever you experience in your daily life, processing the stresses you experience in a more calming way.

Throughout part 2 of the book, you've learned many techniques for influencing the circuitry in your amygdala and keeping it calm. Now it's time to turn to the cortex, which is also capable of initiating, exacerbating, or reducing anxiety. As you've seen in this chapter, when it comes to reducing anxiety, exercise and sleep can benefit both the cortex and the amygdala. In part 3 of the book, we'll take a close look at other ways to take control of your cortex-based anxiety.

PART 3

Taking Control of Your Cortex-Based Anxiety

CHAPTER 10

Thinking Patterns That Cause Anxiety

People tend to treat their emotions as though they have no control over them. But as you're learning, you can influence the underlying neurological processes that give rise to anxiety. In part 2 of the book, we took an in-depth look at how to influence and rewire the amygdala. Here, in part 3, we'll do the same for the cortex. It is possible to change the thoughts, images, and behaviors produced by the cortex, and you can do so in ways that will give you more control over cortex-based anxiety.

Many people are familiar with the idea of using thoughts to control anxiety, having learned about this approach either from therapists or by reading about how thoughts, or cognitions, contribute to creating anxiety. Many more resources are available to assist people in using cortex-based approaches than amygdala-based approaches, and we list a number of books that are helpful in this regard in the Resources section. For now, it's essential that you understand how different approaches can help you rewire your cortex, so you'll know what you want to accomplish when using these techniques. Our goal is not to explain every cortex-based approach in detail, but to show you how these strategies contribute to the process of rewiring your cortex to ease anxiety in lasting ways.

As mentioned in the introduction, "cognition" is the psychological term for cortex processes that most people refer to as "thinking." Perhaps the best-known pioneers of cognitive treatment are psychiatrist Aaron Beck and psychologist Albert Ellis, who each proposed that anxiety can be created or worsened by certain types of thinking. Both suggested that anxiety results from the way people interpret events and sometimes distort reality as a result of certain thinking processes. For example, you may overemphasize the dangerousness of a situation, such as fearing a plane crash despite the overall safety of air travel. Or you may interpret someone else's behavior as being personally relevant when it has nothing to do with you, such as assuming that someone is talking during your presentation because you're boring. Cognitions can cause us to anticipate problems that will never occur or worry about bodily sensations that are quite harmless.

Cognitive Restructuring

The idea underlying the approach known as *cognitive therapy* is that some cognitions are illogical or unhealthy and can create or exacerbate unhealthy patterns of behavior or mental states. Cognitive therapists focus on identifying and changing thoughts that are self-defeating or dysfunctional, particularly thoughts that lead to increased levels of anxiety or depression. This approach is known as *cognitive restructuring*. Cognitive restructuring to combat anxiety intervenes directly in the cortex pathway. When cognitive therapists discuss self-defeating or dysfunctional thoughts, they're focused on processes that occur in the cortex, primarily in the left hemisphere. Of course, whenever we try to change our thoughts, we're trying to modify the cortex in some way. Our thoughts are not simply a result of neurological and chemical processes in the brain; they *are* the neurological and chemical processes in the brain. In cognitive restructuring, the thoughts you think are used to rewire your brain.

As you've been learning, the processes in the brain that create fear and anxiety can and often do occur without involvement of the cortex. Indeed, via the amygdala pathway, fear responses can be put into action before cortex processing has been completed. However, this doesn't mean that thoughts and interpretations don't matter.

They definitely have an impact. It's crucial to have a clear understanding of the ways in which thoughts can affect the amygdala's reactions and the ways in which their impact is limited.

Because anxiety can occur automatically, without input from the cortex's cognitive processing, changing thoughts can't always prevent anxiety. However, when thoughts or images in the cortex have *initiated* the anxiety response, changing those thoughts or images can definitely ease or prevent anxiety. Consider two teenagers waiting for their driver's exams to be scored. Jose sat worrying about whether he'd passed, doubting his answers, and imagining being told that he couldn't get his license. Meanwhile, Ricardo's father joked around with him after he took his exam, which kept Ricardo from focusing on possibly failing. Thanks to his father's distracting antics, Ricardo didn't think about potential negative outcomes. As it turned out, both passed the test, but only Jose had endured a stressful, anxious waiting period. When people change their thoughts, they may be able to prevent cortex-based processes from contributing to their anxiety.

Cognitive restructuring strategies also have the potential to limit amygdala-based anxiety. Often, the cortex *worsens* anxiety initiated by the amygdala. But, rather than adding fuel to the fire, you can learn to control what you're imagining, thinking, or telling yourself and remain more even keeled. As difficult as it may seem to change thoughts and thought processes, it's easier than coping with the emotional reactions created by the amygdala in response to anxiety-provoking thoughts. If you understand the connection between your cortex-based thoughts and activation of your amygdala and recognize the amount of anxiety you can avoid by changing your thoughts, you'll be motivated to work on using your cortex to resist anxiety. And this work has lasting results. By changing your thoughts, you can establish new patterns of responding in the brain that become stable and enduring.

The Power of Interpretations

In chapter 3, we discussed how the cortex's interpretations can increase anxiety. When you experience a situation or event, the situation or event itself doesn't cause you to have an emotion. Despite the fact that people frequently say things like "My husband makes me so mad," it isn't their spouse that causes the emotional reaction. The cortex's *interpretation* of the situation is what leads to the emotional reaction. For example, the cortex may offer an interpretation like "He should notice what I do right and not focus on my mistakes," which leads to angry feelings. If you doubt this, consider that different people have different emotional reactions to the same event. Therefore, the event can't be the cause of the emotion.

As an example, consider this scenario: Josh doesn't show up for a dinner date with Monique and Jayden. Jayden is furious at Josh and expresses her anger. Monique, on the other hand, isn't too concerned and simply wants to enjoy her time with Jayden, whom she hasn't seen in weeks. The event is the same for both women: Josh doesn't show up. But their interpretations are obviously quite different. Jayden's interpretation might be *Josh should follow through when he says he's coming* or *He doesn't have any respect for us*, leading to her reaction of anger. In contrast, Monique has a different interpretation: *This is an opportunity to spend time alone with my friend Jayden.* Her interpretation doesn't result in feelings of anger. Note that each of these interpretations will lead to a different feeling, demonstrating that it is the *interpretation*, not the situation itself, that causes the specific feeling.

Of course, there are other possible interpretations and they will lead to different feelings. If Jayden felt anxious rather than angry, what interpretations could have led to that emotion? If she felt sad, what interpretations could have led to feelings of sadness? It's important to learn that the interpretation you make in a situation can strongly affect what emotional response occurs. (Diagrams showing the influence of interpretations in these examples are available for download at http://www.newharbinger.com/31137; see the back of the book for information on how to access them.)

By being aware of your interpretations during stressful situations and considering the possibility of modifying them, you can begin to take charge of the emotional reactions your cortex causes. Changing your interpretations won't always be easy, because those interpretations are often shaped by your past experiences and expectations. It can take some work to think through the situation and identify the way you want to interpret it. Also, you may not always want to alter your emotional reactions; sometimes they may be appropriate or useful. However, having the ability to alter your cortex's interpretations can often go a long way toward reducing your anxiety.

Exercise: Changing Your Interpretations to Reduce Anxiety

Recognizing that your interpretation of a situation, rather than the situation itself, is causing anxiety gives you a new way to reduce your anxiety. You can use a cortex-based approach and change your interpretations to reduce amygdala activation.

Let's say that Liz is experiencing anxiety about writing assignments in her English class. As in figure 5 (in chapter 3), three elements are at play here: the event, the interpretation provided by Liz's cortex, and her emotion (anxiety). When Liz got a recent writing assignment back, she saw that her teacher had written many comments on the paper. She thought to herself, *All of those comments are pointing out my mistakes. I'm obviously a terrible writer, and I'm going to fail this course.* Immediately after having these thoughts, Liz felt nauseous, started trembling, and felt overwhelmed. Her thoughts had definitely activated her amygdala.

But later, when Liz actually looked at her teacher's comments, she saw that while some of them were indeed corrections, others were compliments, helpful feedback, or her teacher's reactions to thought-provoking things she had written. Her grade was a B—not a disaster, but allowing room for improvement. Now Liz has an opportunity to change her interpretation. Next time she gets a paper back with comments written on it, she can think, *My teacher is giving me helpful feedback. I'm going to learn how to be a better writer and I can get a better grade.* Clearly, these interpretations of the same event won't create the same level of anxiety.

The situations in which you feel anxiety can provide opportunities for you to examine the interpretations your cortex is providing. Keep the three elements in mind: event, interpretation, and resulting emotion. Learn to recognize your interpretations, and then consider how to modify them to reduce anxiety.

Try it now: On a separate piece of paper, list several situations in which you feel anxiety. Then, for each, see if you can identify the interpretations that lead you to react in an anxious manner. (If you have difficulty with this, the assessments later this chapter will be helpful. The items you check on those assessments will provide some insight into the types of interprotations you need to recognize and modify.)

Next, spend some time brainstorming alternative interpretations for each anxiety-igniting interpretation you identified. If you play with this a bit, you can probably see how different interpretations could lead to a wide range of emotional responses. Of course, for the purposes of reducing anxiety, you'd want to focus on interpretations that lead to a more calm, balanced state of mind. (If you need help coming up with alternative interpretations, the section on coping thoughts in chapter 11 will be helpful.)

Once you've identified alternative interpretations, we recommend that you say them out loud in order to establish them more fully. This will strengthen your ability to modify your interpretation. In the beginning, the process of changing interpretations may feel awkward; you may not find your new interpretations convincing. But with time, you'll find that these thoughts become stronger and arise on their own more often. The more you deliberately use them, the more they'll become a part of your habitual way of responding. Remember, the cortex operates on "survival of the busiest" (Schwartz and Begley 2003, 17).

Changing your thoughts isn't easy, but if you devote some attention to noticing your interpretations and are dedicated to looking at situations differently, you can do it. It's worth the effort, since changing your thoughts before your amygdala is activated is much easier than calming yourself down once your amygdala gets involved.

Identifying How Your Own Cortex Initiates Anxiety

In the rest of this chapter, we'll examine several common types of thoughts that frequently activate the amygdala. Learning to recognize them in a variety of situations is an important step in using cognitive restructuring techniques and mindfulness (both discussed in chapter 11) to reduce your anxiety. If you change your thoughts, you establish new patterns of responding in the brain that endure and protect you against anxiety.

Because anxiety-producing thoughts arise automatically, you may not be aware of the various ways in which the cortex creates anxiety. In the following pages, we offer a series of exercises that will help you identify cortex-based processes that contribute to your own anxiety. Note that these assessments aren't professionally designed tests; they're simply offered to help you consider the nature of your own thought processes. As you complete each assessment, consider the examples carefully and be honest about whether they reflect your experience of your own anxiety.

We call all of the cortex-based tendencies below *anxiety-igniting thoughts* because they have the potential to activate the amygdala. In fact, they could be a primary source of your anxiety.

Exercise: Assessing Your Pessimistic Tendencies

One of the simplest ways to see the influence of your cortex is to consider your general outlook on yourself, the world, and the future. Part of the cortex's job is to help you interpret your experiences and make predictions about what's likely to happen in the future. Your general perspective can have a strong impact on this process. Some people tend to be optimistic and expect the best, while others are more pessimistic and expect the worst. Optimism is more common, and it tends to result in less anxiety. If you tend to be pessimistic, you're likely to be more anxious. Furthermore, a pessimistic attitude can make you less willing to try to change your anxiety because you don't expect success.

This assessment will help you examine whether you tend to engage in negative, pessimistic thinking. Read through the statements below and check any that apply to you:

_____ When I have an upcoming presentation or examination, I worry about it quite a bit and fear I won't do well.

_____ I generally expect that if something can go wrong, it will.

_____ I'm often convinced that my anxiety will never end.

_____ When I hear that something unexpected has happened to someone, I typically imagine it's something negative.

_____ I frequently prepare myself for negative events that I fear will occur but seldom or never do.

_____ If it weren't for bad luck, I wouldn't have any luck at all.

_____ Some people want to improve their lives, but that seems pretty hopeless to me.

_____ Most people will let you down, so it's best not to expect much.

If you checked many of these statements, you show signs of pessimistic thinking.

Optimism vs. Pessimism in the Cortex

Optimism is more associated with left hemisphere activation, whereas pessimism is associated with the right hemisphere (Hecht 2013). The right hemisphere is more focused on identifying threats and what can go wrong, so increased activation in the right hemisphere is associated with more negative evaluation. Deliberately attempting to take a positive view of a situation has been shown to activate the left hemisphere (McRae et al. 2012), which is evidence that a pessimistic attitude can be modified.

The *nucleus accumbens*, a structure in the frontal lobes, also plays a role in this. The nucleus accumbens is a pleasure center in the brain that's involved in hope, optimism, and the anticipation of rewards. It's where the neurotransmitter dopamine is released, and studies have shown that when brain levels of dopamine are higher, negative expectations are reduced and optimism increases (Sharot et al. 2012). Neuroscientist Richard Davidson has found that the more activity there is in a person's nucleus accumbens, the more positive outlook the person has (Davidson and Begley 2012). Davidson maintains that this part of the brain underlies optimistic approaches to life, and some researchers have indeed found that the nucleus accumbens reacts differently in pessimists than in optimists (Leknes et al. 2011). Other researchers have found that optimists are more likely to have more activation in the *anterior cingulate cortex*, a brain structure in the frontal lobe (Sharot 2011).

Regardless of whether we can identify specific areas of the cortex where optimistic versus pessimistic tendencies develop, it's clear that pessimism can be modified—and that it's well worth working on. People who are optimistic tend to be happier, handle adversity better, and have better health (Peters et al. 2010). They are more motivated to try to do things and, when they fail, to try again, since they expect something good to come of their efforts (Sharot 2011). They tend to worry less and to focus on positive results, whether or not their expectations are valid.

Conversely, pessimism is more likely to lead to discouragement, withdrawal, and giving up. Pessimists are more likely to worry, imagine undesirable outcomes, and fixate on hardships in life. Focusing on the negative simply isn't an emotionally rewarding way to live. If pessimism is a factor for you, you'll benefit from the cortex-based interventions discussed in chapter 11, including thought-stopping, cognitive restructuring, coping statements, and mindfulness.

Exercise: Assessing Your Tendency to Worry

Worry is a source of anxiety for many people, and the central difficulty for those with generalized anxiety disorder. Worry can involve either images or thoughts. It has a focus on problem solving designed to come up with responses to expected future difficulties. If you have a habit of frequently thinking about possible negative events that could occur, worry may play a role in your anxiety.

This assessment will help you explore whether you tend to worry. Read through the statements below and check any that apply to you:

_____ I'm good at imagining all kinds of things that could go wrong in specific situations.

_____ I sometimes worry that my symptoms are the result of some medical illness that hasn't been diagnosed yet.

_____ I know I tend to worry about trivial things.

_____ When I'm busy at work or with other activities, I don't have as much anxiety.

_____ Even when things are going well, I seem to think about what could go wrong.

_____ I sometimes feel that if I don't worry about specific situations, something will surely go wrong.

_____ If there's even a small possibility that something negative could happen, I tend to dwell on that possibility.

_____ I have trouble falling asleep because of the things I worry about.

If you checked many of these statements, you have a tendency to worry.

Worry Circuits in the Cortex

Anxiety isn't always caused by what's actually happening in our lives. Because of the cortex's ability to anticipate, anxiety can develop from cortex-based thoughts about events that haven't happened and may never happen. Worry is basically thoughts about negative outcomes that could potentially occur. As mentioned, it can involve images or thoughts, and often involves problem-solving thoughts designed to prevent or minimize anticipated future difficulties. Ironically, these attempts to solve problems that may not even occur can create a great deal of distress by fueling anxiety. As nineteenth-century politician and scientist John Lubbock (2004, 188) noted, "A day of worry is more exhausting than a week of work."

Worry largely arises in the *orbitofrontal cortex*, a portion of the frontal lobes that lies above and behind the eyes. This is the brain structure that considers various possible outcomes, both good and bad, and makes decisions about how to act in future situations (Grupe and Nitschke 2013). The orbitofrontal cortex gives us the capacity to plan and exhibit self-control, allowing us to prepare for future events in ways that other animals can't. But our ability to consider different potential outcomes and make decisions based on predictions is a double-edged sword. It can help us anticipate what will happen so we can meet deadlines, prepare dinner on time, and plan our careers. But when anticipation and decision making manifest as worry, we focus primarily on potential negative outcomes and begin to imagine or consider events that are highly unlikely. Some researchers suggest that worrying is a way of trying to use left hemisphere verbal processing to avoid negative images from the right hemisphere (Compton et al. 2008).

A second part of the prefrontal cortex, the anterior cingulate cortex, is also involved in creating worry. It lies in one of the older parts of the prefrontal cortex, and because it's near the center of the brain, it serves as a bridge between the cortex and the amygdala and helps us process emotional reactions in the brain (Silton et al. 2011). Sometimes the anterior cingulate cortex can be overactive, perhaps because of flaws in the way it developed or due to levels of certain neurotransmitters. Instead of relaying information about thoughts and emotions back and forth between the cortex and amygdala like it should and shifting smoothly from one idea to another, it can get stuck on certain ideas or images. The ongoing flow of information back and forth between the frontal cortex and the amygdala, which would allow more flexibility in thinking and responding, gets caught in a loop. When this happens, people become preoccupied with solving potential problems that haven't even developed. We call this a "worry circuit" in the prefrontal cortex. This is quite different from effective planning or problem solving. If you have difficulties with worry, you'll benefit from strategies discussed in chapter 11, including distraction, thought stopping, cognitive restructuring techniques, mindfulness, and learning to plan rather than worry.

Exercise: Assessing Your Tendency Toward Obsessions or Compulsions

As discussed in chapter 3, obsessions involve being preoccupied with a particular situation and unable to stop thinking about it. Compulsions, or repeatedly engaging in specific behaviors, may offer temporary relief, but because these aren't genuinely effective solutions, the need to perform them arises again and again, often in an escalating cycle. If you find yourself preoccupied with certain thoughts or stuck in performing certain compulsions, this is definitely a problem that arises from the cortex pathway.

This assessment will help you identify whether difficulties with obsessions or holding on to thoughts is an issue for you. Read through the statements below and check any that apply to you:

_____ *I can spend a long time rehashing certain events in my mind.*

_____ *When I make some kind of mistake or forget to do something, it takes me a long time to come to terms with it.*

_____ *If a friend or relative disappoints me, it can take months for me to get over being upset and get back on good terms with the person.*

_____ *I tend to get very upset if I can't keep certain objects in order or in good condition.*

_____ *I can become preoccupied with arranging, counting, or evening up things.*

_____ I need to repeatedly check on things in order to reduce my anxiety, either by checking with people or inspecting something, like my stove.

_____ In many situations I just can't stop thinking about the risk of contamination, germs, chemicals, or illness.

_____ Unpleasant thoughts or images frequently come to my mind, and I can't get them out.

If you agreed with many of these statements, obsessive thinking may be a source of your anxiety.

Obsession and Compulsion: Holding on to Certain Thoughts or Behaviors

The cortex can increase your anxiety when it won't let go of a certain idea or behavior. When this happens, you'll feel preoccupied with a particular thought and be unable to stop thinking it, and you may find yourself repetitively engaging in behaviors designed to counteract the thought. Consider Juanita, who had an obsession about her hands possibly being contaminated with dirt or germs. She couldn't stop thinking about what might be on her hands. She also had a compulsion to wash her hands repeatedly and for lengthy periods of time, to the point that they were cracked and bleeding. The problem was, within minutes of washing her hands, she began to fear that she'd become contaminated again, forcing her to wash her hands again.

Although several areas of the cortex may be involved in obsessions, they appear to be associated with activation in the same areas of the cortex involved in worry: the orbitofrontal cortex and the anterior cingulate cortex. The circuitry connecting these two areas has also been the focus of investigation (Ping et al. 2013). Many neuroimaging studies show excessive activation in the orbitofrontal cortex among people with obsessive-compulsive disorder (Menzies et al. 2008). However, this dysfunction need not be permanent.

Research has shown that cognitive behavioral therapy can help reduce obsessive symptoms (Zurowski et al. 2012), and that reduction in symptoms is associated with changes in activation in the orbitofrontal cortex (Busatto et al. 2000).

In regard to the involvement of the anterior cingulate cortex, this part of the brain is supposed to help us shift smoothly between different approaches to responding to problems. But as mentioned in the discussion of worry, sometimes it seems to get caught in a loop. Research suggests that obsessive-compulsive disorder may be due to structural problems in the anterior cingulate cortex, which tends to be thinner in people with obsessive-compulsive disorder (Kuhn et al. 2013).

Obsessive thoughts and compulsive behaviors are both cortex-based processes that contribute to anxiety. Obsessive thoughts tend to focus on certain themes, including contamination, danger, violence, or orderliness, and can create a great deal of anxiety. Compulsions can take various forms but often involve cleaning, checking, counting, or touching. The compulsion itself may not seem to create much anxiety, but when people try to resist their compulsions, they typically experience a great deal of anxiety. In chapter 11, we'll discuss how you can help your cortex resist obsessions. Compulsions may require exposure therapy, described in chapter 8, because resisting them usually activates the amygdala.

Exercise: Assessing Your Perfectionistic Tendencies

Placing unrealistically high standards on yourself or others is guaranteed to increase your anxiety. Because no one is capable of perfection, high standards often mean you're setting yourself up for failure.

This assessment will help you determine whether perfectionism might be an issue for you. Read through the statements below and check any that apply to you:

_____ I have high standards for myself and usually hold myself to them.

_____ I usually have a right way to do something and find it difficult to vary from that approach.

_____ *People consider me extremely conscientious and careful as a worker.*

_____ *When I'm wrong, I'm very embarrassed and ashamed.*

_____ *When others are watching me, I'm concerned that I'll humiliate myself.*

_____ *I almost never perform at a level that I'm satisfied with.*

_____ *I have a hard time letting go of mistakes I make.*

_____ *I feel I have to be hard on myself or I won't be good enough.*

If you checked many of these statements, you may have difficulties with perfectionism.

The Dangers of Perfectionism

Anxiety can arise via the cortex pathway as a result of perfectionist expectations of yourself or others. Sometimes it's clear that people learned perfectionism from others, often their parents. Parents may not see the downside to encouraging their children to always do their best. However, this can create unrealistic expectations in the cortex. This isn't to say parents shouldn't have high expectations for their children, just that they need to be cautious about instilling unrealistic ideas. We simply can't be at our best every moment.

However, parents aren't always the source of perfectionistic tendencies. Consider Tiffany, who recognized that she was the source of her own unrealistic, perfectionistic expectations. She remembered that even as a child, she always felt she had to do everything correctly. Her parents, on the other hand, were more accepting and reasonable. They frequently reassured her that her performance was fine and she didn't need to be perfect.

Whether or not people feel their perfectionistic expectations are reasonable, it's essential to recognize that they're a source of anxiety. The self-criticism and disappointment that result from perfectionism can markedly increase your daily experience of anxiety.

You'll benefit from looking at your expectations that lead to anxiety, because perfectionism may be at their root. Fortunately, the cortex is capable of setting more reasonable expectations, and anxiety will decrease as a result.

Exercise: Assessing Your Tendency to Catastrophize

Catastrophizing is a tendency to see minor problems or small set-backs as huge disasters. If you feel that your whole day is ruined if one specific thing goes wrong, you're catastrophizing. This cortex-based interpretation can result in a great deal of anxiety, but once you learn to recognize it, you can take steps to reduce it.

This assessment will help you determine whether catastrophizing may play a role in your anxiety. Read through the statements below and check any that apply to you:

_____ *I often imagine the worst when I'm thinking about how some situation might turn out.*

_____ *I can make a mountain out of a molehill.*

_____ *People would think I'm going crazy if they knew the awful thoughts that go through my mind.*

_____ *I often feel as if I can't handle one more thing going wrong.*

_____ *When something doesn't turn out the way that I want it to, I find it difficult to cope.*

_____ *I overreact to problems that others wouldn't consider so much of a concern.*

_____ *Even a small setback, like being stopped by a traffic light, can infuriate me.*

_____ *Sometimes what begins as a small doubt in my mind becomes an overwhelming negative thought as I dwell on it.*

If you checked many of these statements, you have a tendency to catastrophize.

The Costs of Overestimating the Costs

If you react to inconveniences as if they're disasters or feel that your whole day is ruined if one minor thing goes wrong, this is definitely heightening your anxiety. Catastrophizing has its roots in the circuitry in the orbitofrontal cortex, which is also involved in worry and, as noted earlier, in considering different outcomes. Another task of the orbitofrontal cortex is estimating the costs or downsides of events (Grupe and Nitschke 2013).

Some people have a tendency to overestimate the costs of certain negative events. For example, when Jeremy is running late and is stopped by a traffic light, he issues a string of obscenities and pounds the steering wheel in fury. Of course, being stopped at the light only adds one or two minutes to his trip, but in his brain, that minor amount of time seems like a cost that warrants the amount of anger and frustration he experiences.

A tendency to react to minor events as if they're going to have disastrous results will definitely activate your amygdala. Ironically, that does increase the costs by adding anxiety to the situation. But tendencies like this can be recognized by catching yourself in the act and replacing catastrophic thoughts with more reasonable coping statements, as suggested in chapter 11.

Exercise: Assessing Your Tendency to Experience Guilt and Shame

Guilt and shame are emotions that come from the frontal and temporal lobes of the cortex. Guilt involves a feeling that you've behaved in a way that you consider to be unacceptable. Shame, on the other hand, is related to feeling that other people will perceive you in a negative way. Both emotions are very anxiety provoking.

This assessment will help you determine whether guilt or shame is an issue for you. Read through the statements below and check any that apply to you:

_____ I frequently feel that I'm not measuring up to what I expect of myself.

_____ I become very concerned when I contemplate not doing something that I feel I should do.

_____ I frequently worry about disappointing people and have trouble saying no.

_____ If a friend is upset when I don't come to an event, I may feel guilty for days.

_____ It feels awful to know I've let someone down.

_____ It's easy for others to guilt-trip me into doing what they want.

_____ It's very hard for me to admit my mistakes and discuss them with others.

_____ Once a person criticizes me, I tend to avoid spending much time around that person.

If you checked many of these statements, guilt, shame, or both are probably contributing to your anxiety.

Guilt, Shame, and Anxiety

As mentioned, guilt involves a feeling that you've behaved in a way that you find unacceptable or that violates a personal standard. Shame is related to feeling that others will perceive you in a negative way. So guilt is focused on your evaluation of yourself, whereas shame involves imagining how others evaluate you. However, both seem to be associated with activation in the frontal and temporal lobes.

Shame and guilt are often involved in social anxiety disorder, which is one of the most common types of anxiety and often involves a fear of being scrutinized by others. Consider Raj, who has difficulty speaking in groups. He tends to feel ashamed, embarrassed, and uncomfortable about how he presents himself, and he expects others to judge him harshly. The truth is, he typically judges himself more harshly than others would, and he also feels guilty about even minor transgressions.

Experiencing high levels of guilt and shame leads to a great deal of anxiety. The amygdala seems to be more strongly activated by shame than by guilt (Pulcu et al. 2014), a finding that's consistent with the role of the amygdala in protecting us from dangers, including the disapproval of others. Cognitive restructuring, including the use of coping thoughts, can slowly change a tendency to respond with guilt and shame.

Exercise: Assessing Right Hemisphere–Based Anxiety

You'll recall that the right hemisphere of the cortex allows you to use your imagination to visualize events that aren't actually occurring. And when you imagine distressing situations, this often inadvertently initiates an anxiety response.

This assessment will help you determine whether the right hemisphere is typically a source of your anxiety. Read through the statements below and check any that you experience often:

_____ *I picture potential problem situations in my mind, imagining various ways things could go wrong and how others will react.*

_____ *I'm very attuned to the tone of people's voices.*

_____ *I can almost always imagine several scenarios that illustrate how a situation could turn out badly for me.*

_____ *I tend to imagine ways that people will criticize or reject me.*

_____ *I often imagine ways that I might embarrass myself.*

_____ *I sometimes see images of terrible events occurring.*

_____ *I rely on my intuition to know what others are feeling and thinking.*

_____ *I'm watchful of people's body language and pick up on subtle cues.*

If you checked many of the statements above, your anxiety may be increased by a tendency to imagine frightening scenarios or rely on intuitive interpretations of people's thoughts that may not be accurate.

Right Hemisphere–Based Anxiety

The right hemisphere specializes in processing experiences in more holistic, integrated ways and is adept at processing nonverbal aspects of human interactions. Sometimes its focus on facial expressions, tone of voice, or body language may cause you to jump to conclusions about this information. For example, you can misinterpret a tone of voice and assume someone is angry or disappointed with you, when they are simply tired.

The right hemisphere has a tendency to focus on negative information, whether that information is visual or auditory (Hecht 2013). We already noted that it tends to be the source of pessimistic thinking. In addition, it can use its powers of imagination to produce scenarios and imagery that can be extremely frightening. The right hemisphere is on the lookout for anything negative in others' posture, tone of voice, or facial expressions.

These right hemisphere–based processes can cause your amygdala to respond as if you're in a dangerous situation when no threat exists. A variety of strategies, including play, meditation, and exercise, can be useful for increasing activation of the left hemisphere, producing positive emotions, and quieting the right hemisphere. We explained these strategies in chapters 6 and 9.

The right hemisphere of the cortex is more active during both anxious arousal and sadness (Papousek, Schulter, and Lang 2009). One study showed that in people with social phobia who were preparing to give a speech, the right side of the brain became activated and heart rate increased (Papousek, Schulter, and Lang 2009).

Neuroscientists have found that the middle portion of the right hemisphere contains an integrated system for responding to immediate threats; this system directs attention to visual scanning of the environment, increases sensitivity to meaningful nonverbal cues, and promotes sympathetic nervous system activity (Engels et al. 2007). This system is always involved once anxiety has begun. However, it can also be engaged when it isn't necessary, in which case it creates anxiety, rather than helping you respond effectively to threats.

In chapter 11, we'll explain how you can use positive imagery from the right hemisphere to combat anxiety. You can also use the melodic and emotional aspects of music, which are processed by the right hemisphere, to engage this hemisphere in positive emotions. In these ways, you can learn to use your right hemisphere to resist anxiety, rather than create it.

Your Personal Profile of Anxiety-Igniting Thoughts

Review the assessments in this chapter. This will give you an overall view of the types of anxiety-producing thoughts you tend to experience, which will help you target your efforts for change. You can't change thoughts you aren't aware of, but once you've identified your problem areas, you can be vigilant for the types of thoughts that most frequently contribute to your cortex-based anxiety. (For a visual representation of your anxiety-igniting thoughts, complete the Anxiety-Igniting Thoughts Profile, which is available for download at http://www.newharbinger.com/31137; see the back of the book for information on how to access it.)

Summary

The assessments in this chapter helped you determine which cortex-influenced processes and thought patterns may be activating your amygdala. Each person has a unique cortex, with its own unique ways of initiating anxiety. Begin watching for your own anxiety-igniting thoughts regularly in your daily life. Being aware of them is the first step in changing them. It's helpful to know which tendencies are your strongest so you can target them specifically. None of these tendencies are fixed and unchangeable. You can rewire your cortex to reduce any of these types of thoughts and strengthen different circuitry to promote alternative processes. In the next chapter, you'll learn a number of techniques that will help you rewire your cortex to ease or resist anxiety.

CHAPTER 11

How to Calm Your Cortex

As you've learned, if you create and dwell on certain thoughts and images in your cortex, you're likely to activate the amygdala and create anxiety. Fortunately, there's a huge difference between thoughts about events and the events themselves. Just because you think about or imagine something occurring doesn't mean it will occur. This difference between your thoughts and external reality is essential to remember because your amygdala may not recognize the distinction. So keep it uppermost in your cortex to help prevent your amygdala from responding to imagined thoughts and images with an anxiety response!

Revisiting Cognitive Fusion

You can gain a great deal of cortex-based control over your anxiety if you recognize the difference between thoughts about events and the events themselves. As discussed in chapter 3, cognitive fusion occurs when we get so caught up in our thoughts that we forget they're merely thoughts. Consider Sonia, a young mother with a baby boy. One day she had a thought about how vulnerable her baby was and how easily she could harm him. Then her mind seemed to fill up with thoughts and images of different ways that she could intentionally or unintentionally hurt her baby. She imagined herself accidentally dropping him and thought about how easily she could drown him. These thoughts and images terrified her, and before long she was afraid to be alone with her son because she believed that having those awful thoughts meant she might act on them. In this way, she confused her thoughts with reality and fell victim to cognitive fusion. Yet the very fact that she was afraid to be alone with her son demonstrated that she was concerned about his being harmed and would take action to protect him if it was necessary.

At any given time, we each have a variety of thoughts created by the cortex, but this doesn't mean that the thoughts are true, that whatever we're thinking about is going to happen, or that we're going to act on our thoughts. Still, it's all too easy to forget that thoughts are just thoughts: neural events in the cortex that may have no relationship to reality. Recognizing the difference between thoughts and actual events is essential in managing cortex-based anxiety.

Exercise: Assessing Your Tendency to Experience Cognitive Fusion

If you have a tendency to take your thoughts and feelings at face value and believe them, this is likely to interfere with your ability to rewire your cortex to help you resist anxiety. The cortex has a great deal of flexibility, but you have to be willing to take advantage of it.

To assess your tendencies toward cognitive fusion, take a moment to read through the statements below and check any that apply to you:

_____ If I don't worry, I'm afraid things will get worse.

_____ When a thought occurs to me, I find I need to take it seriously.

_____ Anxiety is usually a clear sign that something is about to go wrong.

_____ Worrying about something can sometimes prevent bad things from happening.

_____ When I feel ill, I need to focus on it and evaluate it.

_____ I'm afraid of some of my thoughts.

_____ When someone suggests a different way to see things, I have a hard time taking it seriously.

_____ *If I have doubts, there are usually good reasons for them.*

_____ *The negative things I think about myself are probably true.*

_____ *When I expect to do poorly, it usually means I will do poorly.*

If you checked many of these statements, you're probably overly fused with your thoughts and feelings. You'll benefit from recognizing that just thinking or feeling something doesn't make it so. When you believe a thought represents some kind of truth, you'll have more resistance to letting go of that thought, and this can prevent you from rewiring your cortex.

Beware of Cognitive Fusion

Cognitive fusion is quite common. We all tend to assume that what we think is reality, and don't often question our assumptions and interpretations. But sometimes people need to question their perspectives, especially in regard to distressing situations. Knowing that our assumptions are fallible is an important recognition. Cognitive fusion can generate a great deal of unnecessary anxiety.

Cognitive fusion makes people more likely to respond to the *thought* of an event in the same way they'd react if the event actually occurred. Consider Arrianna, who had trouble contacting her boyfriend one afternoon and began to worry that something bad had happened to him. She had images of him being in an accident and also thoughts that he was contemplating breaking up with her. As she considered these possibilities, she became very upset. Later, Arrianna found out that her boyfriend had left his cell phone at home and hadn't received her messages. This was a huge relief to her. What's interesting in this story is that Arrianna reacted to the thoughts she was having as if they were actual events, and those thoughts made her anxious. Do you ever catch yourself doing something similar?

When certain anxiety-igniting thoughts are combined with cognitive fusion, the risk of creating anxiety becomes greater. If you have a tendency to have pessimistic thoughts or to worry, you'll benefit from resisting cognitive fusion. For instance, if you tend to be a pessimistic thinker, it can be helpful to remind yourself that your thoughts don't determine what happens.

We recommend that you examine your own anxiety experiences for evidence of cognitive fusion—accepting thoughts or feelings as true even though there's no evidence, or only weak evidence, to support them. A common example is believing a situation is dangerous because of a *feeling* that it's dangerous, rather than having actual evidence of a threat. Take some time now to make a list of examples of situations where you may be engaging in cognitive fusion. Here are some examples to get you going: *I think my neighbors criticize my lawn*, *Nobody at this party likes me*, or *I absolutely cannot bear to have another panic attack*. Once you've compiled your list, review it and consider how a belief in these unfounded thoughts may be contributing to your anxiety.

Because the amygdala responds to thoughts just as it does to actual events, you may be able to greatly reduce your anxiety by being aware of anxiety-igniting thoughts and reducing the time you spend contemplating such thoughts. Although this sounds logical, surprising numbers of people worry that they must take every thought or feeling they have seriously, and some even argue that the mere existence of a thought suggests it's true, as these examples show:

- An insecure woman insisted that the fact that she didn't have confidence in herself was proof that she shouldn't have confidence in herself.

- An elderly man reported that his fear of falling meant he couldn't leave his home.

- A woman was critical of her work performance and worried that she would be fired—despite never having received a bad evaluation at work.

The cortex is a busy, noisy place, often full of ideas and feelings that have no basis in reality. The problem isn't the ideas and feelings themselves, but a tendency to take them seriously. Psychologist

Steven Hayes (2004, 17) has suggested that "it is the tendency to take these experiences literally and then to fight against them that is...most harmful" and offers cognitive defusion as the solution. Cognitive defusion involves taking a different stance toward your thoughts: being aware of them without getting caught up in them.

Cognitive defusion is a very powerful cognitive restructuring technique. Developing your ability to relate to your thoughts in this way involves not allowing yourself to take thoughts at face value and instead simply recognizing them as experiences you're having. For instance, you could acknowledge a thought without buying into it by saying, "Hmm...interesting. Once again I see that I'm having the thought that I'm never going to get my diploma." To be successful at cognitive defusion, you need to develop a sense of yourself that doesn't get lost in the thought processes of your cortex. You're an observer of your cortex, not a believer of everything it produces. To help distance yourself from a thought, you could tell yourself something like, "I need to be careful of this pesky thought. I have no reason to put faith in it, and it's likely to activate my amygdala." Mindfulness techniques, which we'll discuss later in this chapter, are also very helpful, as they help you build strength and skill in focusing your thoughts on what you choose and resisting the urge to get lost in thoughts that may or may not reflect reality.

Develop Some Healthy Skepticism About Your Cortex

In many ways, your cortex creates the world you live in, processing your sensations and allowing you to perceive and think about your experiences. It also allows you to reflect on past experiences and imagine the future. This can make it difficult to remember that the information you experience in your cortex isn't the same as reality. For example, you may think what you saw during a robbery was completely accurate, but we know from court trials that eyewitness accounts are notoriously erroneous. Sometimes even our eyes play tricks on us, and this can be true of our other senses too. We see the world through our cortex, but there is much more going on than we are aware of (such as ultraviolet light, high- and low-frequency sounds, or other people's private thoughts). At http://www.newhar binger.com/31137, you'll find a downloadable PowerPoint presentation that illustrates how the cortex can make you perceive something that doesn't exist, keep you from perceiving what actually is there, or cause you to think something makes perfect sense when it's actually nonsense. We encourage you to take a look at this information. (See the back of the book for information on how to access the presentation.)

Controlling Your Anxiety-Igniting Thoughts

At this point, you might want to review the assessments you completed in chapter 10 to identify your most common anxiety-igniting thoughts and target them for change. If your cortex is producing such thoughts, don't let it run rampant. You can change the thoughts in your cortex and shift your focus to other thoughts. This lays the groundwork for changing the circuitry in your cortex. In the remainder of this chapter, we'll describe cognitive restructuring techniques you can use to do that. Because there are so many excellent self-help books entirely devoted to this topic, we won't give you detailed instructions for all of the strategies. If you find the approaches we describe helpful, we highly recommend that you check out some of the books listed in the Resources section at the end of the book.

Cognitive restructuring techniques give you the power to literally change your cortex. The key is to be skeptical of anxiety-igniting thoughts and dispute them with evidence, ignore them as if they don't exist, or replace them with new, more adaptive thoughts, also known as coping thoughts. Pay particular attention to the anxiety-igniting thoughts you catch yourself using quite often. Remember, neural circuitry is strengthened by the principle of "survival of the busiest" (Schwartz and Begley 2003, 17), so the more you think certain thoughts, the stronger they become. If you interrupt anxiety-provoking thoughts and images and repeatedly replace them with new cognitions, you can literally change the circuitry of your brain.

Using Coping Thoughts

Coping thoughts are thoughts or statements that are likely to have positive effects on your emotional state. One way of evaluating the usefulness of thoughts is to look at the effects they have on you. In this light, you can clearly see the value of coping thoughts, which are more likely to result in calm responding and an increased ability to cope with difficult situations. Here are some examples.

Anxiety-Igniting Thought	Coping Thought
It's no use trying. Things will never work out for me.	*I'm going to try, because then there's at least a chance that I'll accomplish something.*
Something's going to go wrong. I can feel it.	*I don't know what's going to happen. These kinds of feelings have been wrong before.*
I need to focus on this thought, doubt, or concern.	*Cortex, you've spent too much time on this and need to move on.*
I must be competent and excel at everything I do.	*No one is perfect. I'm human and expect I'll make mistakes at times.*
Everyone should like me.	*No one is liked by everyone, so I'll encounter people who don't like me.*
I can't stand this!	*This isn't the end of the world. I'll survive.*
I can't help worrying about this.	*Worrying never fixes anything. It only upsets me.*

I don't want to disappoint other people.	Trying to please everyone is impossible and stresses me out. Let it go.
I can't handle this situation.	I'm a competent person, and even though I don't like this situation, I can get through it.

Of course, you'll have to be vigilant about recognizing anxiety-igniting thoughts and substituting coping thoughts, but it's worth the effort. Some people post their coping thoughts to remind themselves. By deliberately thinking coping thoughts at every possible opportunity, you can rewire your cortex to produce coping thoughts on its own. Remember, you're changing your neural circuitry!

Be sure to focus on the types of thoughts that are most problematic for you. Consult your Anxiety-Igniting Thoughts Profile (mentioned at the end of chapter 10) if you downloaded it and filled it out. For example, if you tend toward perfectionism, it's useful to watch for "musts" and "shoulds" in your thinking. When you tell yourself you "must" accomplish something or that something "should" happen according to a certain plan or schedule, you're setting yourself up for stress and worry. The words "must" and "should" make it seem like a rule is being violated if your performance is less than perfect or events don't unfold as planned. If nothing else, replace "I should..." with "I'd like to..." That way, you aren't creating a rule that must be followed. Instead, you're simply expressing a goal or a desire—one that may or may not be met. It's a kinder, gentler thought.

Replacing Thoughts (Because You Can't Erase Them)

When people work on changing thoughts, they often complain that they can't get rid of their negative thoughts. This is a common problem that springs from how the mind works. Studies have shown that trying to erase or silence a thought simply isn't an effective approach (Wegner et al. 1987). For example, if you're asked to not think about pink elephants, the image of pink elephants will, of course, leap into your mind, even if you haven't been thinking about pink elephants all day. And the harder you try to stop thinking about pink elephants, the more you think of them. If you have a tendency toward obsession, you're probably familiar with this pattern. Erasing a thought by constantly reminding yourself not to think about it (and therefore thinking about it) activates the circuitry storing that thought and makes it stronger.

You might be successful in interrupting a thought by specifically telling yourself "Stop!" This technique is called *thought stopping*. However, the next step is crucial. If you *replace* the thought with another thought, it's more likely that you'll keep the first thought out of your mind. Let's say you're working in your garden and keep worrying that at any moment you'll encounter a snake. Tell yourself "Stop!" and then begin thinking about something else: a song on the radio, the names of the flowers you intend to plant in your garden, ideas you have for a loved one's birthday present—basically anything captivating and, ideally, pleasant. By replacing the anxiety-provoking thought with something else that engages your mind, you make it more likely that you won't return to that thought.

Therefore, "Don't erase—replace!" is the best approach with anxiety-igniting thoughts. If you notice that you're thinking something like, *I can't handle this*, focus on replacing that thought with a coping thought, such as *This isn't easy, but I will get through it*. By repeating this coping thought to yourself, you'll strengthen a more adaptive way of thinking and activate circuitry that will protect you from anxiety. It takes some practice, but your new thoughts will eventually become habitual.

Changing the Anxiety Channel

Some people have a strong tendency to use the cortex in ways that create anxiety. They are often quite talented at imagining dreadful events or coming up with negative scenarios. In fact, people who are highly creative and imaginative are sometimes more prone to anxiety for this very reason. The way they think about their life and imagine events frequently captures the attention of the amygdala and provokes a reaction. People who catastrophize or use their right brain imagery in ways that frighten them are typical examples.

If this is an issue for you, think of your cortex as cable television. Despite having hundreds of channels to choose from, you get stuck on the Anxiety Channel. Unfortunately, it appears to be your favorite. You may focus on thoughts and images that have anxiety-igniting potential without realizing it. Or perhaps you're aware of this focus but argue with the thoughts, just as you might argue with televised political commentators you don't agree with. Arguing with your thoughts is similar. You don't want to spend too much time arguing with your thoughts because that tends to keep the focus on them and maintain the circuitry underlying them.

Consider Rachel, who recently had a job interview. At the time, she felt the interview went fairly well; but afterward she started rethinking some of her statements and wondering how they sounded to the person who interviewed her. Now, with each passing day Rachel feels increasingly worried about whether she'll get the position. She becomes discouraged and starts to worry she won't get the job. She begins to second-guess how she responded in the interview, becomes pessimistic, and starts to believe she won't get the job. Rachel is definitely watching the Anxiety Channel.

Notice that the interview isn't Rachel's real problem. She doesn't even know how the interview affected her chances of being hired. The Anxiety Channel is the problem. If Rachel recognizes this and, instead of focusing on worrying about the interview, begins to look at other job possibilities and prepare herself for new interviews, she'll be much more productive. If she imagines future interviews going better because of what she learned from this interview, her attitude will be much more positive. As Rachel begins to think about strategies for upcoming interviews, she finds she's no longer stuck on the Anxiety Channel.

Rachel changed the channel by shifting her focus from the past to the future, but there are many ways to change the channel. One way is through *distraction*: moving your focus of attention to something completely different. Distraction can be a very effective way to manage anxiety. For example, instead of thinking about the stress of an upcoming dentist appointment, change the channel and focus on a different topic. You could focus on having a conversation with someone, coming up with menus for the week, or playing with your kids or a pet. Distracting yourself by focusing on other activities or ideas is one of the simplest ways to change the channel.

One of the best kinds of distraction is play. So many anxious people are gripped by an excruciating seriousness and therefore have difficulty loosening up and having fun. Cultivating a sense of playfulness is essential. And it isn't necessary to wait until you aren't anxious to become playful. Be playful to find relief. Playing games, joking, and engaging in silliness are some of the best distractions. Humor is essential in coping with life's challenges.

Using distraction to change the channel can immediately reduce anxiety in a given situation. But beyond that, the more you deliberately direct your attention to other topics when you notice you're focused on anxiety-igniting thoughts, the more you increase activity in new circuits and reduce activity in circuits focused on anxiety-producing topics or images. The circuitry that you use the most becomes the strongest, and circuitry you don't use becomes weaker and less likely to be activated. So you don't just reduce your anxiety for a few moments; you rewire your cortex.

Replacing Worry with Planning

Worry may be one of the most seductive cortex-based processes. For people who tend to worry, it often feels helpful to think about a problem, concern, or responsibility and invest time in anticipating potential difficulties. But if constantly focusing on your concerns tends to be self-perpetuating and activates your amygdala, is it really helping?

As discussed in chapter 10, it can be easy to get stuck in worry, imagining one negative event after another and considering endless possible responses. You may worry about events long before it's even necessary to prepare for them and waste time deciding how to respond to imagined events that may never occur. Researchers have shown that when people continue to think about a negative event, they lengthen their emotional reaction to the event, maintaining negative emotions for longer than they otherwise would have lasted (Verduyn, Van Mechelen, and Tuerlinckx 2011).

Instead of getting stuck in worrying or ruminating, plan! If you anticipate that a situation will actually arise, come up with possible solutions and then move on to other thoughts. If the situation actually arises, you can put your plan in place. In the meanwhile, you don't need to keep thinking about it.

Here's an example: Anne's son Joey had a birthday coming up, and Anne heard that her aunt Janice would be attending his birthday party. Anne recalled a recent argument with Janice and began to worry that another argument would occur. She then got stuck in thoughts about potential conflicts with her aunt, imagining various criticisms Janice could raise and considering how she might respond. She worried about what Janice might say about her to others at the party and started thinking of ways to respond to other people who could become involved. But luckily, Anne had been down this route before and realized that her worries about how to deal with her aunt were actually producing more anxiety. She recognized that her tendency to worry was making her anticipate a big scene that might not even happen. She told herself "Stop!" and said to herself, "My plan is to get ready for the party. I'll deal with Janice later—if I need to."

When the day of the party came, Anne's aunt seemed primarily focused on little Joey, and her conversations with Anne related to events going on in her own children's lives. In the end, Anne's recognition of her tendency to worry and her decision to interrupt it and make a plan saved her a great deal of unnecessary anxiety.

Considering Medications

Certain medications can be helpful as you attempt to change your thoughts. As discussed in chapter 8, new circuitry is less likely to be generated if you're taking benzodiazepines, which may explain why multiple studies have found that the people who benefit most from therapy are those who aren't taking benzodiazepines (for example, Addis et al. 2006 and Ahmed, Westra, and Stewart 2008). In contrast, certain medications, including selective serotonin reuptake inhibitors (SSRIs) and serotonin-norepinephrine reuptake inhibitors (SNRIs) can be very helpful for people who are having difficulty changing their thinking patterns, because these medications promote the development of new circuitry.

A gardening analogy might be useful at this point. Taking SSRIs and SNRIs is similar to using fertilizer in a garden to promote new growth. You'll see more roots, branches, and buds. Of course, you need to be careful what you fertilize, because weeds will also respond to fertilizer, perhaps even more quickly. Similarly, it's important to be very deliberate about which neural patterns you're strengthening to make the most effective use of SSRI or SNRI treatment. You need to consider what you're teaching your cortex when you take these medicines. They are most helpful in changing thought processes when people also engage in therapy that's focused on modifying problematic thoughts (Wilkinson and Goodyer 2008). Remember, if you'd like to learn more about various antianxiety medications and when they may or may not be helpful, you'll find a bonus chapter on this topic at http://www.newharbinger.com/31137. (See the back of the book for information on how to access it.)

Attending to the Right Hemisphere

If the right hemisphere is a source of your anxiety, rewiring your cortex so that you use the left hemisphere more often can be helpful. The right hemisphere specializes in negative emotions and avoidance, whereas the left hemisphere has more focus on approaching what a person is interested in (Davidson 2004), so increasing activity in your left hemisphere will be beneficial. Seek out activities that engage the left hemisphere, such as watching amusing programs, reading thought-provoking articles, playing games, and exercise. All of these activities can reduce the dominance of right hemisphere–based reactivity. Meditation has also been shown to increase left hemisphere activity, and we'll address this topic shortly, when we discuss mindfulness.

Another approach is to deliberately engage the right hemisphere in an activity that's incompatible with negative mood states. Listening to uplifting music is a good example. For nonmusicians, music is primarily processed on the right side of the brain. (Learning to perform music brings in more left hemisphere skills.) When you listen to music you enjoy, you directly engage your right hemisphere in positive emotional responding. You might also consider singing, which activates the right hemisphere more than speaking (Jeffries, Fritz, and Braun 2003). Deliberately using music to improve your mood, increase your energy level, and replace negative thinking is a wonderful right hemisphere approach to resisting anxiety.

Positive forms of imagery are another way to engage the right hemisphere in activity that's incompatible with anxiety. When you use your imagination to take yourself to a pleasant location and imagine it in elaborate sensory detail, as described in chapter 6, you're working your right hemisphere. So imagine a positive scene, with all the sights, sounds, smells, and physical sensations your right hemisphere can provide. It's an excellent and inexpensive vacation from anxiety.

Using the Power of Mindfulness

Anxiety has the ability to hijack your cortex, dominate your conscious awareness, and take over your life. But what if you could find a way to use the cortex to look at your anxiety, seeing it from a distance rather than living in it and being trapped by its influence? What if you could use your cortex to get outside the anxiety so it's just an experience you're having? Mindfulness is a cortex-based technique that does exactly that.

Mindfulness is an age-old approach that's been practiced in various traditions for thousands of years. Therefore, it's been described and defined in many ways. Psychiatrist Jeffrey Brantley describes mindfulness as a friendly acceptance and deep awareness of your current experience; in his book *Calming Your Anxious Mind* (2007), he explains how the simple skill of mindful awareness can defeat anxiety. Our natural responses to anxiety are to try to escape it or control it, or to get caught up in suffering it. But mindfulness gives you another path, with origins in Eastern practices of meditation—an approach of being open to and accepting of whatever you're feeling. In this approach, as psychologist Steven Hayes (2004, 9) puts it, "A 'negative thought' mindfully observed won't necessarily have a negative function." You might think of this approach as training your cortex to lovingly, patiently observe your anxiety responses, much like a caring, patient parent might observe a child's temper tantrum—closely noting all aspects of the behavior and remaining loving and nonreactive until the child calms down.

In essence, mindfulness means understanding that all you ever really have is the present moment, and practicing a new way to inhabit and observe that moment: with a focus on allowing, accepting, and being fully aware of whatever you're experiencing. This may sound simple, but it takes practice. However, this practice can be woven into your life. You can transform your typical daily experiences into opportunities to practice mindfulness as you eat breakfast, listen to the sounds in your yard, focus on walking, or concentrate on a deep breathing practice. You'll soon see how different these experiences feel when you mindfully attend to them. You'll also realize how often you're caught up in thoughts that keep you from truly experiencing life. For example, one woman reported that when she began mindfulness practice, she realized she hadn't really tasted her breakfast for years. As she established a practice of beginning her day mindfully while eating, she found that it set a very different tone for her day.

After you learn to focus on mindfully observing fairly neutral everyday experiences, you can turn your awareness to your anxiety. Through practice, you relax your body and train your cortex to take on a nonjudging attitude, an openness to what's happening that puts you in the role of peaceful, detached observer, rather than someone who's struggling with anxiety and its physical symptoms.

Exercise: A Mindful Approach to Anxiety

The next time you feel anxiety, seek a quiet place to practice mindfulness. Let your focus be on your bodily experience, and allow your awareness of anything else to fade. If your attention wanders, simply bring it back to the experience of anxiety in your body. For example, if you feel the flush of adrenaline, consider the experience and simply allow yourself to feel it. How intense is it? What parts of your body are affected? What sensations do you have? How do the sensations change over time? Look at your body to see if you notice signs of anxiety. Are you trembling? Are your legs trying to move? Also notice the impulses you have, perhaps to say something or to leave. Be aware of these impulses without acting on them, and observe what happens to them as you observe. Likewise, notice the thoughts that are coming into your mind. You don't have to analyze them; just let them be there. Don't judge yourself as you make these observations; simply observe. Accept your anxiety as a normal process. Let yourself experience it as it moves through you, changing over time, without fighting it or encouraging it. Simply observe.

Try to practice mindfulness in response to anxiety for about a month, whenever you can take time to attend to your anxiety. You can further develop your ability to use mindfulness with anxiety by focusing on different components of your anxiety response. For example, one time you might choose to focus on how your breathing is affected, another time on your heart, another on your thoughts, and so on. Notice how your sense of your anxiety changes when you take this approach.

Control Might Not Be the Answer

In this book, you've learned that the cortex has limited ways of exerting direct control over amygdala-based responding once it occurs. But the truth is, you don't have to control the amygdala's responding if you use mindfulness to observe it without getting caught up in it. When you take a mindful approach to the anxiety response, the cortex gives up the goal of controlling the situation and simply allows anxiety to happen. This acceptance of your experience is the ultimate antidote to anxiety.

Much of the power of anxiety comes from the constant struggle to fight it and make it stop; that's how it can exert so much control over your life. When you face the experience of anxiety, knowing that it will pass and accepting it, it will actually pass more quickly. You won't perpetuate it with a fearful reaction to it. Much of the discomfort of anxiety arises from struggling with it and trying to wish it away. Strange as it may seem, by giving up attempts to control anxiety, you can actually be more in control of your brain.

Studies show amazing changes in the brains of people who practice mindfulness and other forms of meditation. In addition to being able to reduce their anxiety in the present moment (Zeidan et al. 2013), they experience lasting changes in the cortex that make them resistant to anxiety. Those who are experienced in mindfulness haven't changed the amygdala's responding; they've disengaged the cortex from getting caught up in the amygdala's responding (Froeliger et al. 2012). With mindfulness, you train the cortex to respond to anxiety in a completely new way. Neuroimaging studies show that the few parts of the cortex that have a direct connection to the amygdala—the ventral medial prefrontal cortex and the anterior cingulate cortex—are the very parts of the cortex activated by mindfulness meditation (Zeidan et al. 2013). These findings indicate that mindfulness approaches can help you rewire parts of the cortex that are intimately connected with calming anxiety.

The ultimate power of mindfulness training is that it changes the way your cortex responds to anxiety. Making mindfulness part of your daily life is the best way to use it to transform your anxiety. We highly recommend that you explore mindfulness in depth. There are many excellent books and other resources that provide training in mindfulness, and some of them have a specific focus on anxiety. (See the Resources section at the back of the book for some recommendations.)

Summary

In this chapter, we explained several approaches to helping your cortex respond to anxiety in new ways. As you use these approaches to rewire your cortex, you'll become increasingly able to live the way you want to live. Perhaps most importantly, you've learned that in addition to reducing and preventing anxiety, you can use mindfulness to help your cortex accept anxiety. All of these techniques can help you live a more anxiety-resistant life. The final step is to put everything you've learned in this book together; and in the conclusion, we'll help you do just that.

CONCLUSION

Putting It All Together to Live an Anxiety-Resistant Life

We hope this book has given you an awareness of the brain processes involved in anxiety, and that what you've learned will help you live your life the way you want to. Understanding how anxiety is created in the amygdala and the way in which the cortex pathway contributes to anxiety helps you understand that your anxiety isn't completely within your conscious control. You can't change the fact that your brain is designed to produce the experience of anxiety. But you can learn to cope with anxiety. Furthermore, the neuroplasticity of the brain, which has been demonstrated in a number of studies, opens the door to rewiring your brain to alter your experience of anxiety.

Even though aspects of anxiety are beyond your conscious control, that doesn't mean anxiety has to control your life. No one will ever live a life entirely free of anxiety, but we can all diminish anxiety's effects on our lives by using both amygdala-based and cortex-based strategies.

Your new understanding of the role of the amygdala and the influences of the cortex is valuable knowledge that will help you identify the sources of your anxiety. You can use this information to specifically target the processes underlying your anxiety, allowing you to set realistic goals and make lasting changes in your brain. You now know how to create new connections in your cortex by practicing new ways of thinking and interpreting until they become habitual. You've also learned about the power and potential of mindfulness and acceptance. You know how to rewire your amygdala by providing it with new experiences that stimulate it to make new connections. Once an anxiety reaction begins and it's too late to stop it, you know how to choose amygdala-based strategies to limit its impact and cortex-based strategies that will help you to let go of the desire to control it.

Where to Start

Because you've learned so many strategies in this book, you may be wondering where to begin. The best way to start is by focusing on calming your amygdala. Begin with relaxation. Learn the skills of slowing your breathing and relaxing your muscles in order to turn off your sympathetic nervous system and activate your parasympathetic nervous system, as discussed in chapter 6. Also use positive imagery, exercise, sleep, and music to calm your amygdala, as outlined in chapters 6, 9, and 11. Practice relaxation strategies repeatedly, every day, to lower your overall anxiety level, integrating various kinds of relaxation into your life until it becomes second nature to relax. All of these approaches will lead to fairly rapid changes in your amygdala's daily functioning.

Next, focus your attention on cortex-based strategies as needed. Review chapter 10 to remind yourself of the types of anxiety-igniting thoughts that are most problematic for you, and use the approaches described in chapter 11 to combat those thoughts. Practice monitoring and modifying your thoughts until you're able to think in more productive and anxiety-resistant ways in most situations. You might also consider whether certain medications may be helpful in this process.

Keep in mind the life goals that are important to you, perhaps revisiting the exercise at the end of the introduction from time to time to remind yourself of your goals or identify new ones. Then be on the lookout for situations where anxiety keeps you from pursuing your goals. Helping you achieve those goals is our ultimate aim in this book. To take charge of your life, identify triggers for anxiety in situations where anxiety or compulsions are blocking your goals, as discussed in chapter 7. Then target those triggers with exposure, as outlined in chapter 8, to reduce the limiting effects of anxiety. Use exposure for each problematic trigger situation until you feel your fear decrease, signaling that rewiring has occurred in your amygdala.

When you feel stressed by the exposure exercises, remind yourself that you need to activate your amygdala in order for it to learn. You can't make new connections unless you experience some anxiety, so you have to activate to generate. As you begin to experience less anxiety about triggers that are blocking your goals, you'll feel more in control of your life. The process of rewiring your brain to reduce anxiety will be gradual, but your brain will adapt to the experiences you provide and the patterns of thinking you cultivate, and it will build new circuitry. Although you'll experience some setbacks, you'll gradually see an improvement in your ability to take charge of your life as you use these strategies. To quickly recap, here's the sequence we recommend:

1. Use relaxation, sleep, and exercise to reduce sympathetic nervous system activation.

2. Monitor your thinking for any anxiety-igniting thoughts.

3. Replace anxiety-igniting thoughts with coping thoughts.

4. Determine your life goals and what interferes with those goals.

5. Identify triggers of fear and anxiety that interfere with your goals.

6. Design exposure exercises that can modify your amygdala's response to these triggers.

7. Practice exposure exercises until you notice a decrease in your anxiety and fear.

Strengthening Your Resolve

Although the approach outlined in this book may seem overwhelming, if you break it down into steps, you'll find it quite manageable. You'll see improvements with each step, and that will encourage you. When you see that you're able to use the strategies in chapter 6 to relax, you'll feel more confident about managing your anxiety. When you experience beneficial changes in your thinking as a result of the approaches in chapter 11, you'll be heartened. And when you feel how exposure reduces your anxiety, you'll become increasingly able to push through your fears.

Throughout, it's important to remember that your ultimate goal is to rewire your brain, so with each step, try to keep in mind what's happening in your brain. Every strategy you use sends an important message to your brain; with repetition, your brain will adapt. Don't be daunted by the prospect of ongoing practice. After all, this is necessary to excel in almost any endeavor, from arithmetic to athletics. You're taking charge of your life, step-by-step. Of course, there will be challenges along the way. The following pointers may be helpful when you need to strengthen your resolve.

Act Despite Your Anxiety

Taking action in the face of fear is easier said than done. But that is exactly what's needed to transform your experience of anxiety and rewire your brain. Remember that courage is to act despite your fear.

You've learned so much about anxiety in this book. It's a complicated, multifaceted experience based on complex neurological processes. You're likely to encounter people who won't understand even a fraction of what you now know about anxiety. Don't allow their judgments to discourage you. You may face more feelings of terror before lunch than most people do all year. The people in your life may not recognize that, for you, getting to home base means running six or seven bases, not just four. But if you realize that and give yourself credit for what you're doing, it can help a great deal. Your friends may not know that going out with them is an exhausting accomplishment, not a lighthearted evening. Recognize what you're able achieve while contending with anxiety, and take pride in it.

Take It One Day—or One Minute— at a Time

We encourage you to take life one day at a time. In everyday practice, this means living in the present moment and not focusing on worries about what may or may not happen in the future. By keeping your focus on the present moment, you'll save your mental energy for the tasks in front of you. Plus, why would you want to linger on the Anxiety Channel, reliving stressful events of the past and imagining frightening future scenarios? You're likely to miss some of the best experiences in your life if you stay focused on the Anxiety Channel.

In moments of stress, it can be immensely helpful to zero in on just one minute at a time. Sometimes getting through a specific moment is all we can handle—and all we need to handle. It's perfectly reasonable to focus on coping with one situation at a time. Luckily, life is presented to us one minute at a time—actually, one second at a time. All we truly need to do is get through each minute, especially when confronting anxiety. Sometimes, getting through just a few minutes is an achievement in and of itself. Taking life one minute at a time can sometimes make life easier to handle.

Focus on the Positive

Your life is made up of countless varying moments. If you can learn to focus your brain on positive experiences and savor them, you'll feel generally happier. Tune in to the moments filled with joy and beauty when they come and hold on to these experiences. Cultivate playfulness. Cherish those you love. Ultimately, love is stronger than fear.

Setbacks will come in life, but they're often simply a sign that you're testing the limits. Of course ships are safe in a harbor, but they aren't meant to stay there. If you never have setbacks, you probably aren't setting your sights very high. In any case, it isn't necessary to dwell on setbacks. You can find beauty and pleasure in life if you look for them. Consciously and mindfully experience every happy event and feel the delight that you get out of these special moments. How you focus your thoughts has a very powerful influence on your brain. Focus your brain on the positive, beautiful, and enjoyable aspects of your life. You'll be happier as a result.

Don't Mind Your Anxiety

Whether you were born with a brain that tends to create anxiety or acquired your anxiety problems as a result of life experiences, you can cope with anxiety. Even if the anxiety pathways in your brain are activated, you can use the approaches in this book to change your responses and, with time, rewire your brain. The key is to focus on the positive and not let anxiety control you. All of the knowledge you've gained in this book will help you manage your anxiety more effectively and gradually rewire your brain to reduce your experience of anxiety. We hope this journey brings you relief, encouragement, and joy. You deserve it!

Resources

As mentioned in chapters 10 and 11, there are numerous self-help books outlining cognitive behavioral and mindfulness approaches to easing anxiety through cortex-based methods. Here are a few we recommend.

Cognitive Behavioral Approaches

Anxiety and Avoidance: A Universal Treatment for Anxiety, Panic, and Fear, by Michael Tompkins

The Anxiety and Worry Workbook: The Cognitive Behavioral Solution, by David Clark and Aaron Beck

The Cognitive Behavioral Workbook for Anxiety: A Step-by-Step Program, by William J. Knaus

The PTSD Workbook: Simple, Effective Techniques for Overcoming Traumatic Stress, by Mary Beth Williams and Soili Poijula

Prisoners of Belief: Exposing and Changing Beliefs That Control Your Life, by Matthew McKay and Patrick Fanning

Stop Obsessing! How to Overcome Your Obsessions and Compulsions, by Edna Foa and Reid Wilson

When Perfect Isn't Good Enough: Strategies for Coping with Perfectionism, by Martin Antony and Richard Swinson

Mindfulness Approaches

Calming the Rush of Panic: A Mindfulness-Based Stress Reduction Guide to Freeing Yourself from Panic Attacks and Living a Vital Life, by Bob Stahl and Wendy Millstine

Calming Your Anxious Mind: How Mindfulness and Compassion Can Free You from Anxiety, Fear, and Panic, by Jeffrey Brantley

The Mindfulness and Acceptance Workbook for Anxiety: A Guide to Breaking Free from Anxiety, Phobias, and Worry Using Acceptance and Commitment Therapy, by Georg Eifert and John Forsyth

A Mindfulness-Based Stress Reduction Workbook for Anxiety, by Bob Stahl, Florence Meleo-Meyer, and Lynn Koerbel

The Mindfulness Code: Keys for Overcoming Stress, Anxiety, Fear, and Unhappiness, by Donald Altman

Mindfulness for Beginners: Reclaiming the Present Moment—and Your Life, by Jon Kabat-Zinn

The Mindfulness Workbook for OCD: A Guide to Overcoming Obsessions and Compulsions Using Mindfulness and Cognitive Behavioral Therapy, by Jon Hershfield and Tom Corboy

The Mindful Path Through Worry and Rumination: Letting Go of Anxious and Depressive Thoughts, by Sameet Kumar

The Mindful Way Through Anxiety: Break Free from Chronic Worry and Reclaim Your Life, by Susan Orsillo and Lizabeth Roemer

Things Might Go Terribly, Horribly Wrong: A Guide to Life Liberated from Anxiety, by Kelly Wilson and Troy DuFrene

The Worry Trap: How to Free Yourself from Worry and Anxiety Using Acceptance and Commitment Therapy, by Chad LeJeune

References

Addis, M. E., C. Hatgis, E. Cardemile, K. Jacob, A. D. Krasnow, and A. Mansfield. 2006. "Effectiveness of Cognitive-Behavioral Treatment for Panic Disorder Versus Treatment as Usual in a Managed Care Setting: 2-Year Follow-Up." *Journal of Consulting and Clinical Psychology* 74:377–385.

Ahmed, M., H. A. Westra, and S. H. Stewart. 2008. "A Self-Help Handout for Benzodiazepine Discontinuation Using Cognitive Behavior Therapy." *Cognitive and Behavioral Practice* 15:317–324.

Amano, T., C. T. Unal, and D. Paré. 2010. "Synaptic Correlates of Fear Extinction in the Amygdala." *Nature Neuroscience* 13:489–495.

Anderson, E., and G. Shivakumar. 2013. "Effects of Exercise and Physical Activity on Anxiety." *Frontiers in Psychiatry* 4:1–4.

Armony, J. L., D. Servan-Schreiber, J. D. Cohen, and J. E. LeDoux. 1995. "An Anatomically Constrained Neural Network Model of Fear Conditioning." *Behavioral Neuroscience* 109:246–257.

Barad, M. G., and S. Saxena. 2005. "Neurobiology of Extinction: A Mechanism Underlying Behavior Therapy for Human Anxiety Disorders." *Primary Psychiatry* 12:45–51.

Bequet, F., D. Gomez-Merino, M. Berhelot, and C. Y. Guezennec. 2001. "Exercise-Induced Changes in Brain Glucose and Serotonin

Revealed by Microdialysis in Rat Hippocampus: Effect of Glucose Supplementation." *Acta Physiologica Scandinavica* 173:223–230.

Bourne, E. J., A. Brownstein, and L. Garano. 2004. *Natural Relief for Anxiety: Complementary Strategies for Easing Fear, Panic, and Worry.* Oakland, CA: New Harbinger.

Brantley, J. 2007. *Calming Your Anxious Mind*, 2nd ed. Oakland, CA: New Harbinger.

Broman-Fulks, J. J., and K. M. Storey. 2008. "Evaluation of a Brief Aerobic Exercise Intervention for High Anxiety Sensitivity." *Anxiety, Stress, and Coping* 21:117–128.

Broocks, A., T. Meyer, C. H. Gleiter, U. Hillmer-Vogel, A. George, U. Bartmann, and B. Bandelow. 2001. "Effect of Aerobic Exercise on Behavioral and Neuroendocrine Responses to Meta-chlorophenylpiperazine and to Ipsapirone in Untrained Healthy Subjects." *Psychopharmacology* 155:234–241.

Busatto, G. F., D. R. Zamignani, C. A. Buchpiguel, G. E. Garrido, M. F. Glabus, E. T. Rocha, et al. C. 2000. "A Voxel-Based Investigation of Regional Cerebral Blood Flow Abnormalities in Obsessive-Compulsive Disorder Using Single Photon Emission Computed Tomography (SPECT)." *Psychiatry Research: Neuroimaging* 99:15–27.

Cahill, S. P., M. E. Franklin, and N. C. Feeny. 2006. "Pathological Anxiety: Where We Are and Where We Need to Go." In *Pathological Anxiety: Emotional Processing in Etiology and Treatment*, edited by B. O. Rothbaum. New York: Guilford.

Cain, C. K., A. M. Blouin, and M. Barad. 2003. "Temporally Massed CS Presentations Generate More Fear Extinction Than Spaced Presentations." *Journal of Experimental Psychology: Animal Behavior Processes* 29:323–333.

Cannon, W. B. 1929. *Bodily Changes in Pain, Hunger, Fear, and Rage.* New York: Appleton.

Claparede, E. 1951. "Recognition and 'Me-ness.'" In *Organization and Pathology of Thought*, edited by D. Rapaport. New York: Columbia University Press.

Compton, R. J., J. Carp, L. Chaddock, S. L. Fineman, L. C. Quandt, and J. B. Ratliff. 2008. "Trouble Crossing the Bridge: Altered Interhemispheric Communication of Emotional Images in Anxiety." *Emotion* 8:684–692.

Conn, V. S. 2010. "Depressive Symptom Outcomes of Physical Activity Interventions: Meta-analysis Findings." *Annals of Behavioral Medicine* 39:128–138.

Cotman, C. W., and N. C. Berchtold. 2002. "Exercise: A Behavioral Intervention to Enhance Brain Health and Plasticity." *Trends in Neurosciences* 25:295–301.

Crocker, P. R., and C. Grozelle. 1991. "Reducing Induced State Anxiety: Effects of Acute Aerobic Exercise and Autogenic Relaxation." *Journal of Sports Medicine and Physical Fitness* 31:277–282.

Croston, G. 2012. *The Real Story of Risk: Adventures in a Hazardous World*. Amherst, NY: Prometheus Books.

Davidson, R. J. 2004. "What Does the Prefrontal Cortex 'Do' in Affect: Perspectives on Frontal EEG Asymmetry Research." *Biological Psychology* 67:219–233.

Davidson, R. J., and S. Begley. 2012. *The Emotional Life of Your Brain: How Its Unique Patterns Affect the Way You Think, Feel, and Live—and How You Can Change Them*. New York: Hudson Street Press.

DeBoer L., M. Powers, A. Utschig, M. Otto, and J. Smits. 2012. "Exploring Exercise as an Avenue for the Treatment of Anxiety Disorders." *Expert Review of Neurotherapeutics* 12:1011–1022.

Delgado, M. R., K. I. Nearing, J. E. LeDoux, and E. A. Phelps. 2008. "Neural Circuitry Underlying the Regulation of Conditioned Fear and Its Relation to Extinction." *Neuron* 59:829–838.

Dement, W. C. 1992. *The Sleepwatchers.* Stanford, CA: Stanford Alumni Association.

Desbordes, L. T., T. W. W. Negi, B. A. Pace, C. L. Wallace, C. L. Raison, and E. L. Schwartz. 2012. "Effects of Mindful-Attention and Compassion Meditation Training on Amygdala Response to Emotional Stimuli in an Ordinary, Non-meditative State." *Frontiers in Human Neuroscience* 6, article 292.

Dias, B., S. Banerjee, J. Goodman, and K. Ressler. 2013. "Towards New Approaches to Disorders of Fear and Anxiety." *Current Opinion on Neurobiology* 23:346–352.

Doidge, N. 2007. *The Brain That Changes Itself: Stories of Personal Triumph from the Frontiers of Brain Science.* New York: Penguin.

Drew, M. R., and R. Hen. 2007. "Adult Hippocampal Neurogenesis as Target for the Treatment of Depression." *CNS and Neurological Disorders—Drug Targets* 6:205–218.

Dunn, A. L., T. G. Reigle, S. D. Youngstedt, R. B. Armstrong, and R. K. Dishman. 1996. "Brain Norepinephrine and Metabolites After Treadmill Training and Wheel Running in Rats." *Medicine and Science in Sports and Exercise* 28:204–209.

Dwyer, K. K., and M. M. Davidson. 2012. "Is Public Speaking Really More Feared Than Death?" *Communication Research Reports* 29:99–107.

Engels, A. S., W. Heller, A. Mohanty, J. D. Herrington, M. T. Banich, A. G. Webb, and G. A. Miller. 2007. "Specificity of Regional Brain Activity in Anxiety Types During Emotion Processing." *Psychophysiology* 44:352–363.

Fagard, R. H. 2006. "Exercise Is Good for Your Blood Pressure: Effects of Endurance Training and Resistance Training." *Clinical and Experimental Pharmacology and Physiology* 33:853–856.

Feinstein, J. S., R. Adolphs, A. Damasio, and D. Tranel. 2011. "The Human Amygdala and the Induction and Experience of Fear." *Current Biology* 21:34–38.

Foa, E. B., J. D. Huppert, and S. P. Cahill. 2006. "Emotional Processing Theory: An Update." In *Pathological Anxiety: Emotional Processing in Etiology and Treatment*, edited by B. O. Rothbaum. New York: Guilford.

Froeliger, B. E., E. L. Garland, L. A. Modlin, and F. J. McClernon. 2012. "Neurocognitive Correlates of the Effects of Yoga Meditation Practice on Emotion and Cognition: A Pilot Study." *Frontiers in Integrative Neuroscience* 6:1–11.

Goldin, P. R., and J. J. Gross. 2010. "Effects of Mindfulness-Based Stress Reduction (MBSR) on Emotion Regulation in Social Anxiety Disorder." *Emotion* 10:83–91.

Greenwood, B. N., P. V. Strong, A. B. Loughridge, H. E. Day, P. J. Clark, A. Mika, et al. 2012. "5–HT2C Receptors in the Basolateral Amygdala and Dorsal Striatum Are a Novel Target for the Anxiolytic and Antidepressant Effects of Exercise." *PLoS One* 7:e46118.

Grupe, D. W., and J. B. Nitschke. 2013. "Uncertainty and Anticipation in Anxiety: An Integrated Neurobiological and Psychological Perspective." *Nature Reviews Neuroscience* 14:488–501.

Hale, B. S., and J. S. Raglin. 2002. "State Anxiety Responses to Acute Resistance Training and Step Aerobic Exercise Across Eight Weeks of Training." *Journal of Sports Medicine and Physical Fitness* 42:108–112.

Hayes, S. C. 2004. "Acceptance and Commitment Therapy and the New Behavior Therapies." In *Mindfulness and Acceptance:*

Expanding the Cognitive-Behavioral Tradition, edited by S. C. Hayes, V. M. Follette, and M. M. Linehan. New York: Guilford.

Hebb, D. O. 1949. *The Organization of Behavior.* New York: Wiley.

Hecht, D. 2013. "The Neural Basis of Optimism and Pessimism." *Experimental Neurobiology* 22:173–199.

Heisler, L. K., L. Zhou, P. Bajwa, J. Hsu, and L. H. Tecott. 2007. "Serotonin 5–HT2c Receptors Regulate Anxiety-Like Behavior." *Genes, Brain, and Behavior* 6:491–496.

Hoffmann, P. 1997. "The Endorphin Hypothesis." In *Physical Activity and Mental Health*, edited by W. P. Morgan. Washington, DC: Taylor and Francis.

Jacobson, E. 1938. *Progressive Relaxation.* Chicago: University of Chicago Press.

Jeffries, K. J., J. B. Fritz, and A. R. Braun. 2003. "Words in Melody: An H215O PET Study of Brain Activation During Singing and Speaking." *NeuroReport* 14:749–754.

Jerath, R., V. A. Barnes, D. Dillard-Wright, S. Jerath, and B. Hamilton. 2012. "Dynamic Change of Awareness During Meditation Techniques: Neural and Physiological Correlates." *Frontiers in Human Science* 6:1–4.

Johnsgard, K. W. 2004. *Conquering Depression and Anxiety Through Exercise.* Amherst, NY: Prometheus Books.

Kalyani, B. G., G. Venkatasubramanian, R. Arasappa, N. P. Rao, S. V. Kalmady, R. V. Behere, H. Rao, M. K. Vasudev, and B. N. Gangadhar. 2011. "Neurohemodynamic Correlates of 'Om' Chanting: A Pilot Functional Magnetic Resonance Imaging Study." *International Journal of Yoga* 4:3–6.

Keller, J., J. B. Nitschke, T. Bhargava, P. J. Deldin, J. A. Gergen, G. A. Miller, and W. Heller. 2000. "Neuropsychological Differentiation of Depression and Anxiety." *Journal of Abnormal Psychology* 109:3–10.

Kessler, R. C., W. T. Chiu, O. Demler, and E. E. Walters. 2005. "Prevalence, Severity, and Comorbidity of 12-Month *DSM-IV* Disorders in the National Comorbidity Survey Replication (NCS-R)." *Archives of General Psychiatry* 62:617–627.

Kim, M. J., D. G. Gee, R. A. Loucks, F. C. Davis, and P. J. Whalen. 2011. "Anxiety Dissociates Dorsal and Ventral Medial Prefrontal Cortex Functional Connectivity with the Amygdala at Rest." *Cerebral Cortex* 21:1667–1673.

Kuhn, S., C. Kaufmann, D. Simon, T. Endrass, J. Gallinat, and N. Kathmann. 2013. "Reduced Thickness of Anterior Cingulate Cortex in Obsessive-Compulsive Disorder." *Cortex* 49:2178–2185.

LeDoux, J. E. 1996. *The Emotional Brain: The Mysterious Underpinnings of Emotional Life.* New York. Simon and Schuster.

LeDoux, J. E. 2000. "Emotion Circuits in the Brain." *Annual Review of Neuroscience* 23:155–184.

LeDoux, J. E. 2002. *Synaptic Self: How Our Brains Become Who We Are.* New York: Viking.

LeDoux, J. E., and J. M. Gorman. 2001. "A Call to Action: Overcoming Anxiety Through Active Coping." *American Journal of Psychiatry* 158:1953–1955.

LeDoux, J. E., and D. Schiller. 2009. "The Human Amygdala: Insights from Other Animals." In *The Human Amygdala*, edited by P. J. Whalen and E. A. Phelps. New York: Guilford.

Leknes, S., M. Lee, C. Berna, J. Andersson, and I. Tracey. 2011. "Relief as a Reward: Hedonic and Neural Responses to Safety from Pain." *PLoS One* 6:e17870.

Linden, D. E. 2006. "How Psychotherapy Changes the Brain: The Contribution of Functional Neuroimaging." *Molecular Psychiatry* 11:528–538.

Lubbock, J. 2004. *The Use of Life.* New York: Adamant Media Corporation.

Maron, M., J. M. Hettema, and J. Shlik. 2010. "Advances in Molecular Genetics of Panic Disorder." *Molecular Psychiatry* 15:681–701.

McRae, K., J. J. Gross, J. Weber, E. R. Robertson, P. Sokol-Hessner, R. D. Ray, J. D. Gabrieli, and K. N. Ochsner. 2012. "The Development of Emotion Regulation: An fMRI Study of Cognitive Reappraisal in Children, Adolescents, and Young Adults." *Social Cognitive and Affective Neuroscience* 7:11–22.

Menzies, L., S. R. Chamberlain, A. R. Laird, S. M. Thelen, B. J. Sahakian, and E. T. Bullmore. 2008. "Integrating Evidence from Neuroimaging and Neuropsychological Studies of Obsessive-Compulsive Disorder: The Orbitofronto-Striatal Model Revisited." *Neuroscience and Biobehavioral Reviews* 32:525–549.

Milham, M. P., A. C. Nugent, W. C. Drevets, D. P. Dickstein, E. Leibenluft, M. Ernst, D. Charney, and D. S. Pine. 2005. "Selective Reduction in Amygdala Volume in Pediatric Anxiety Disorders: A Voxel-Based Morphometry Investigation." *Biological Psychiatry* 57:961–966.

Molendijk, M. L., B. A. Bus, P. Spinhoven, B. W. Penninx, G. Kenis, J. Prickaertz, R. C. Voshaar, and B. M. Elzinga. 2011. "Serum Levels of Brain-Derived Neurotrophic Factor in Major Depressive Disorder: State-Trait Issues, Clinical Features, and Pharmacological Treatment." *Molecular Psychiatry* 6:1088–1095.

Nitschke, J. B., W. Heller, and G. A. Miller. 2000. "Anxiety, Stress, and Cortical Brain Function." In *The Neuropsychology of Emotion*, edited by J. C. Borod. New York: Oxford University Press.

Nolen-Hoeksema, S. 2000. "The Role of Rumination in Depressive Disorders and Mixed Anxiety/Depressive Symptoms." *Journal of Abnormal Psychology* 109:504–511.

Ochsner, K. N., R. R. Ray, B. Hughes, K. McRae, J. C. Cooper, J. Weber, J. D. E. Gabrieli, and J. J. Gross. 2009. "Bottom-Up and

Top-Down Processes in Emotion Generation." *Association for Psychological Science* 20:1322–1331.

Ohman, A. 2007. "Face the Beast and Fear the Face: Animal and Social Fears as Prototypes for Evolutionary Analyses of Emotion." *Psychophysiology* 23:125–145.

Ohman, A., and S. Mineka. 2001. "Fears, Phobias, and Preparedness: Toward an Evolved Module of Fear and Fear Learning." *Psychological Review* 108:483–522.

Olsson, A., K. I. Nearing, and E. A. Phelps. 2007. "Learning Fears by Observing Others: The Neural Systems of Social Fear Transmission." *Social Cognitive and Affective Neuroscience* 2:3–11.

Papousek, I., G. Schulter, and B. Lang. 2009. "Effects of Emotionally Contagious Films on Changes in Hemisphere Specific Cognitive Performance." *Emotion* 9:510–519.

Pascual-Leone, A., A. Amedi, F. Fregni, and L. B. Merabet. 2005. "The Plastic Human Brain Cortex." *Annual Review of Neuroscience* 28:377–401.

Pascual-Leone, A., and R. Hamilton. 2001. "The Metamodal Organization of the Brain." *Progress in Brain Research* 134:427–445.

Peters, M. L., I. K. Flink, K. Boersma, and S. J. Linton. 2010. "Manipulating Optimism: Can Imagining a Best Possible Self Be Used to Increase Positive Future Expectancies?" *Journal of Positive Psychology* 5:204–211.

Petruzzello, S. J., and D. M. Landers. 1994. "State Anxiety Reduction and Exercise: Does Hemispheric Activation Reflect Such Changes?" *Medicine and Science in Sports and Exercise* 26:1028–1035.

Petruzzello, S. J., D. M. Landers, B. D. Hatfield, K. A. Kubitz, and W. Salazar. 1991. "A Meta-analysis on the Anxiety-Reducing Effects of Acute and Chronic Exercise: Outcomes and Mechanisms." *Sports Medicine* 11:143–182.

Phelps, E. A. 2009. "The Human Amygdala and the Control of Fear." In *The Human Amygdala*, edited by P. J. Whalen and E. A. Phelps. New York: Guilford.

Phelps, E. A., M. R. Delgado, K. I. Nearing, and J. E. LeDoux. 2004. "Extinction Learning in Humans: Role of the Amygdala and vmPFC." *Neuron* 43:897–905.

Ping, L., L. Su-Fang, H. Hai-Ying, D. Zhange-Ye, L. Jia, G. Zhi-Hua, X. Hong-Fang, Z. Yu-Feng, and L. Zhan-Jiang. 2013. "Abnormal Spontaneous Neural Activity in Obsessive-Compulsive Disorder: A Resting-State Functional Magnetic Resonance Imaging Study." *PLoS One* 8:1–9.

Pulcu, E., K. Lythe, R. Elliott, S. Green, J. Moll, J. F. Deakin, and R. Zahn. 2014. "Increased Amygdala Response to Shame in Remitted Major Depressive Disorder." *PLoS One* 9(1):e86900.

Quirk, G. J., J. C. Repa, and J. E. LeDoux. 1995. "Fear Conditioning Enhances Short-Latency Auditory Responses of Lateral Amygdala Neurons: Parallel Recordings in the Freely Behaving Rat." *Neuron* 15:1029–1039.

Rimmele, U., B. C. Zellweger, B. Marti, R. Seiler, C. Mohiyeddini, U. Ehlert, and M. Heinrichs. 2007. "Trained Men Show Lower Cortisol, Heart Rate, and Psychological Responses to Psychosocial Stress Compared with Untrained Men." *Psychoneuroendocrinology* 32:627–635.

Sapolsky, R. M. 1998. *Why Zebras Don't Get Ulcers: An Updated Guide to Stress, Stress-Related Diseases, and Coping.* New York: W. H. Freeman.

Schmolesky, M. T., D. L. Webb, and R. A. Hansen. 2013. "The Effects of Aerobic Exercise Intensity and Duration on Levels of Brain-Derived Neurotrophic Factor in Healthy Men." *Journal of Sports Science and Medicine* 12: 502–511.

Schwartz, J. M., and S. Begley. 2003. *The Mind and the Brain: Neuroplasticity and the Power of Mental Force.* New York: Harper Collins.

Sharot, T. 2011. "The Optimism Bias." *Current Biology* 21:R941–R945.

Sharot, T., M. Guitart-Masip, C. W. Korn, R. Chowdhury, and R. J. Dolan. 2012. "How Dopamine Enhances an Optimism Bias in Humans." *Current Biology* 22:1477–1481.

Shiotani H., Y. Umegaki, M. Tanaka, M. Kimura, and H. Ando. 2009. "Effects of Aerobic Exercise on the Circadian Rhythm of Heart Rate and Blood Pressure." *Chronobiology International* 26:1636–1646.

Silton R. L., W. Heller, A. S. Engels, D. N. Towers, J. M. Spielberg, J. C. Edgar, et al. 2011. "Depression and Anxious Apprehension Distinguish Frontocingulate Cortical Activity During Top-Down Attentional Control." *Journal of Abnormal Psychology* 120:272–285.

Taub, E., G. Uswatte, D. K. King, D. Morris, J. E. Crago, and A. Chatterjee. 2006. "A Placebo-Controlled Trial of Constraint-Induced Movement Therapy for Upper Extremity After Stroke." *Stroke* 37:1045–1049.

Van der Helm, E., J. Yao, S. Dutt, V. Rao, J. M. Salentin, and M. P. Walker. 2011. "REM Sleep Depotentiates Amygdala Activity to Previous Emotional Experiences." *Current Biology* 21: 2029–2032.

Verduyn, P., I. Van Mechelen, and F. Tuerlinckx. 2011. "The Relation Between Event Processing and the Duration of Emotional Experience." *Emotion* 11:20–28.

Walsh, R., and L. Shapiro. 2006. "The Meeting of Meditative Disciplines and Western Psychology: A Mutually Enriching Dialogue." *American Psychologist* 61:227–239.

Warm, J. S., G. Matthews, and R. Parasuraman. 2009. "Cerebral Hemodynamics and Vigilance Performance." *Military Psychology* 21:75–100.

Wegner, D., D. Schneider, S. Carter, and T. White. 1987. "Paradoxical Effects of Thought Suppression." *Journal of Personality and Social Psychology* 53:5–13.

Wilkinson, P. O., and I. M. Goodyer. 2008. "The Effects of Cognitive-Behaviour Therapy on Mood-Related Ruminative Response Style in Depressed Adolescents." *Child and Adolescent Psychiatry and Mental Health* 2:3–13.

Wilson, R. 2009. *Don't Panic: Taking Control of Anxiety Attacks*, 3rd ed. New York: Harper Perennial.

Wolitzky-Taylor, K. B., J. D. Horowitz, M. B. Powers, and M. J. Telch. 2008. "Psychological Approaches in the Treatment of Specific Phobias: A Meta-analysis." *Clinical Psychology Review* 28:1021–1037.

Yoo, S., N. Gujar, P. Hu, F. A. Jolesz, and M. P. Walker. 2007. "The Human Emotional Brain Without Sleep: A Prefrontal Amygdala Disconnect." *Current Biology* 17:877–878.

Zeidan, F., K. T. Martucci, R. A. Kraft, J. G. McHaffie, and R. C. Coghill. 2013. "Neural Correlates of Mindfulness Meditation–Related Anxiety Relief." *Social Cognitive and Affective Neuroscience* 9:751–759.

Zurowski, B., A. Kordon, W. Weber-Fahr, U. Voderholzer, A. K. Kuelz, T. Freyer, K. Wahl, C. Buchel, and F. Hohagen. 2012. "Relevance of Orbitofrontal Neurochemistry for the Outcome of Cognitive-Behavioural Therapy in Patients with Obsessive-Compulsive Disorder." *European Archives of Psychiatry and Clinical Neuroscience* 262:617–624.

Catherine M. Pittman, PhD, is associate professor at Saint Mary's College in Notre Dame, IN. As a licensed clinical psychologist in private practice in South Bend, IN, she specializes in the treatment of brain injuries and anxiety disorders. She is a member of the Anxiety and Depression Association of America (ADAA), and provides workshops and seminars on the topics of anxiety and stress.

Elizabeth M. Karle, MLIS, is collection management supervisor at the Cushwa-Leighton Library at Saint Mary's College in Notre Dame, IN. In addition to supplying research for this book, she has personal experience with anxiety disorders—providing a first-hand perspective that focuses the book on what is most useful for the anxiety sufferer. Originally from Illinois, she currently resides in South Bend, IN, and holds degrees or certificates from the University of Notre Dame, Roosevelt University, and Dominican University. She is author of *Hosting a Library Mystery*.

Real change *is* possible

For more than forty-five years, New Harbinger has published proven-effective self-help books and pioneering workbooks to help readers of all ages and backgrounds improve mental health and well-being, and achieve lasting personal growth. In addition, our spirituality books offer profound guidance for deepening awareness and cultivating healing, self-discovery, and fulfillment.

Founded by psychologist Matthew McKay and Patrick Fanning, New Harbinger is proud to be an independent, employee-owned company. Our books reflect our core values of integrity, innovation, commitment, sustainability, compassion, and trust. Written by leaders in the field and recommended by therapists worldwide, New Harbinger books are practical, accessible, and provide real tools for real change.

newharbingerpublications

FROM OUR PUBLISHER—

As the publisher at New Harbinger and a clinical psychologist since 1978, I know that emotional problems are best helped with evidence-based therapies. These are the treatments derived from scientific research (randomized controlled trials) that show what works. Whether these treatments are delivered by trained clinicians or found in a self-help book, they are designed to provide you with proven strategies to overcome your problem.

Therapies that aren't evidence-based—whether offered by clinicians or in books—are much less likely to help. In fact, therapies that aren't guided by science may not help you at all. That's why this New Harbinger book is based on scientific evidence that the treatment can relieve emotional pain.

This is important: if this book isn't enough, and you need the help of a skilled therapist, use the following resources to find a clinician trained in the evidence-based protocols appropriate for your problem. And if you need more support—a community that understands what you're going through and can show you ways to cope—resources for that are provided below, as well.

Real help is available for the problems you have been struggling with. The skills you can learn from evidence-based therapies will change your life.

Matthew McKay, PhD
Publisher, New Harbinger Publications

If you need a therapist, the following organization can help you find a therapist trained in cognitive behavioral therapy (CBT).

The Association for Behavioral & Cognitive Therapies (ABCT) Find-a-Therapist service offers a list of therapists schooled in CBT techniques. Therapists listed are licensed professionals who have met the membership requirements of ABCT and who have chosen to appear in the directory.

Please visit www.abct.org and click on *Find a Therapist*.

For additional support for patients, family, and friends, please contact the following:

Anxiety and Depression Association of American (ADAA)
please visit www.adaa.org

MORE BOOKS *from*
NEW HARBINGER PUBLICATIONS

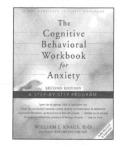

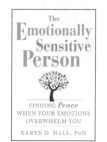

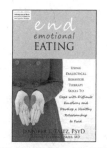

Register your **new harbinger** titles for additional benefits!

When you register your **new harbinger** title—purchased in any format, from any source—you get access to benefits like the following:

- Downloadable accessories like printable worksheets and extra content

- Instructional videos and audio files

- Information about updates, corrections, and new editions

Not every title has accessories, but we're adding new material all the time.

Access free accessories in 3 easy steps:

1. Sign in at NewHarbinger.com (or **register** to create an account).

2. Click on **register a book**. Search for your title and click the **register** button when it appears.

3. Click on the **book cover or title** to go to its details page. Click on **accessories** to view and access files.

That's all there is to it!

If you need help, visit:

NewHarbinger.com/accessories

new harbinger
CELEBRATING
40 YEARS